The Health and Safety Trainer's Guidebook

The book is designed to help trainers design and deliver health and safety training in a fun and high-impact way, such that trainees engage with the subject and remember how to apply it in the work environment. It will be useful for managers, trainers, professionals and graduate students in the fields of ergonomics, human factors, and occupational health and safety.

The text comprehensively explains the effect of the brain on learning and then develops the training processes from training needs analysis all the way through effective training techniques and ending with competence assurance. The unique approach of the book is that Part II provides a range of 30 tried and tested original resource ideas to make health and safety training effective and memorable. It will be a valuable text for professionals and graduate students in the fields of ergonomics, human factors, and occupation health and safety.

- Provides practical and tested solutions to common training problems.
- Covers a resource section showing how to develop interesting and relevant training exercises.
- Focuses on special needs of health and safety training.
- Guides the reader through identifying the training need, delivering the training and finally assuring competence.
- Designed to help trainers design and deliver health and safety training in a fun and high-impact way.

The Health and Safety Trainer's Guidebook

Simon W. Pain

CRC Press is an imprint of the
Taylor & Francis Group, an **informa** business

First edition published 2023
by CRC Press
6000 Broken Sound Parkway NW, Suite 300, Boca Raton,
FL 33487-2742

and by CRC Press
4 Park Square, Milton Park, Abingdon, Oxon, OX14 4RN

CRC Press is an imprint of Taylor & Francis Group, LLC

ISBN: 978-1-032-37958-6 (hbk)
ISBN: 978-1-032-31242-2 (pbk)
ISBN: 978-1-003-34277-9 (ebk)

DOI: 10.1201/9781003342779

Typeset in Sabon
by MPS Limited, Dehradun

For Anne
Remember to work safely, somebody wants you back home!

Contents

PART II

Training resources section 179

Author

Simon W. Pain is a retired multi-award-winning independent Safety, Health and Environmental Management consultant based in Scotland. He has a wealth of health, safety and environmental management experience in various manufacturing industries gained over the last 50 years.

A Chartered Mechanical engineer, with over 32 years of experience in senior management positions with British Steel, ICI and DuPont's engineering, manufacturing, research and corporate functions. He has been advising company executives at the board level on Safety, Health and Management issues for the last 25 years. He spent many years as Divisional Safety, Health and Environmental Manager for ICI and Dupont, the latter being widely regarded as the world benchmark company for health and safety standards. During the last 20 years he has developed novel techniques in Health and Safety training and communication which have been commended by the Institute of Occupational Safety and Health in November 2004.

As a consultant, Simon Pain specialised in raising awareness and motivating senior managers to achieve a paradigm shift in health and safety awareness. He did this by using the high-impact approach and making the subject interesting and fun.

He is an expert in auditing, especially at the management level and personally designed and developed the ICI audit protocol system to ensure that auditing standards were consistent. As a DNV trained auditor he has led audits not only in the United Kingdom, but also in the USA, India, Japan, Malaysia, Belgium, Germany, Spain and the Netherlands. He has extensive experience of lecturing and training in all aspects of health, safety and environmental management.

He is a fellow of both the UK's Institution of Mechanical Engineers and the Energy Institute and a Chartered member of the Institute of Occupational Safety and Health.

He was also a member of the UK Government's Energy Best Practice Committee and a Board Member for the Solway River Purification Board until the formation of the new Scottish Environment Protection Agency.

Simon graduated in Mechanical Engineering from the University of Birmingham and earned his postgraduate qualifications in Health and Safety from the University of Loughborough.

He is the author of two international books on Safety, Health and Environmental auditing, published by CRC Press/Taylor & Francis and has received seven industry awards over the last five years, including International Consultancy of the Year in 2017 and 2021.

Part I

Chapter 1

Introduction

Parents have a wide range of responsibilities to their children. One of those is to educate them and pass on knowledge. It is similar in the work environment. Young inexperienced people are joining the workforce all the time, and, equally, very skilled and knowledgeable older people are leaving, taking a lifetime of experience away with them into retirement, where the skill requirements are very different. It is very easy for the skills and knowledge of the most experienced people in the workforce to get lost when they retire. If society is to continue to function, then we all have a duty to pass on our knowledge and experience to the succeeding generation.

The need for education and training exists in all walks of life, but it is especially important at work. When something goes wrong at work and someone gets injured or there is some sort of adverse environmental impact, there is a natural desire to want to prevent a recurrence. In well-run organisations, these events are usually investigated in order to identify what went wrong, so that we can learn and prevent it from happening again in the future. The experienced investigator quickly learns that one of the many outcomes of this type of investigation is almost invariably that some sort of additional training is recommended. The reality is that we can never really do enough training.

The purpose of this book is to focus on the training that has implications for reducing the risk to people and the environment. Most people that I have met, whether at the management or shop floor level, will say that health and safety is important. The problem is that we all have different views about what is "safe". It is easy to say that something was unsafe after an accident or incident has occurred, but the aim must be to prevent harm from occurring in the first place. It is common to hear the phrase "Oh it's 'elf & safety mate!" as an excuse or an irritation. This is because the two parties involved have different perceptions of what is safe or acceptable. The problem with this agreement that "health & safety is important", is that after that general agreement, the topic has the potential to be very dull or even boring if put across in the wrong way. Effective health and safety

DOI: 10.1201/9781003342779-2

training needs to be challenging without being so complicated that participants cannot follow it and become disengaged.

Health and safety at work is a diverse and complex subject that requires a broad range of skills and knowledge. The breadth of this expertise is continually expanding with the advent of more and more detailed regulations and the effects of burgeoning new technologies and techniques. This expansion of demand presents an ever-increasing requirement for education and training.

The key purpose of this book is to inspire and help those of you who are involved in practical health and safety training to become skilled and passionate about the standards of your training and to ensure that we get the right messages across to the right people at the right time and that those messages are remembered and acted upon.

This book contains 35 well-tried and tested activities and exercises for health and safety training events. These are detailed in the resources section in the second half of the book.

Chapter 2

The importance of memory to learning – How adults learn

Learning for adults is not like being at school, where there is a set curriculum and children are guided through the subjects and information that they need to learn. As adults we never stop learning from things that we see and experience. The problem in a health and safety context is that quite often learning by experience can sometimes have negative consequences and lead to accidents. As adults, we learn best if we recognise for ourselves the need to learn. In this situation we will be much more motivated and receptive to learning. The challenge for the trainer is to stimulate that need.

It is quite possible that the people who are being trained don't really want to be there. Health and safety is important but also potentially uninteresting. The tutor needs to create impact and this is best done by surprising the audience. You want them to go away thinking that "I didn't expect to enjoy that, but it has made me think!"

This next fictitious extract is reproduced with kind permission from A.J. Roberts:

> An elderly man, who's living alone in Scotland wanted to plant his annual tomato garden, but it was very difficult work, since the ground was very hard.

His only son, Paul, who used to help him, is now in prison. The old man wrote this letter to his son and described his predicament:

> *Dear Paul,*
> *I am feeling pretty sad, because it looks like I won't be able to plant my tomato garden this year. I'm just getting too old to be digging up a garden plot. I know if you were here my troubles would be over. I know you would be happy to dig the plot for me, like in the old days.*
> *Love, Dad*

DOI: 10.1201/9781003342779-3

A few days later he received a letter from his son:

> *Dear Dad,*
> *Don't dig up the garden. That's where the bodies are buried!*
> *Love, Paul.*

At 4 a.m. the next morning, the police arrived and dug up the entire area without finding any bodies. They apologised to the old man and left.

That same day the old man received another letter from his son:

> *Dear Dad,*
> *Go ahead and plant the tomatoes now. That's the best I could do under the circumstances.*
> *Love Paul.*

The moral of that story – Always think outside the box to create the outcome and impact that you need. People will remember it!

Psychologists tell us that the brain goes into "auto-shut-off" after as little as 10 minutes. This means that we will need to vary our style, delivery and approach in order to retain the trainee's attention. Trainees can think at a rate of about 800 words/minute, whereas as a tutor at best you will only be able to speak at about 120 words/minute. This means that if you are not being stimulating, your audience will have a surplus 680 words/minute to occupy thinking about the football match, how bored they are or whether they can get home early! However, retaining attention is not the only challenge. The problem with educating people with new information is that unfortunately they often do not remember what has been said. Often, we can recall minute details of things that happened to us years ago but cannot remember what we were taught or told just last week.

I was once asked to act as an expert witness in the case of an individual who worked in a factory as a "service operator". This entailed the operator doing a wide range of cleaning operations. The operator had been dismissed as a result of not wearing suitable personal protective equipment while decanting the chemical residue out of a portable bulk container. It later turned out that the residue was acidic, although the operator had not been told that and in any case, English was not the operator's first language. A week before the dismissal, the operator had attended an hour-long training session about "Chemical Hazards". The individual was dismissed on the basis of "Gross misconduct" as it was claimed that he had only just been trained in the hazards of chemicals. We must remember that being trained does not mean that you have necessarily understood and learned the messages. Training is not the same thing as competence. It transpired later, that the training that the operator had been given, involved watching a video on cleaning hazards, but this video had been produced specifically for the education of cleaners in offices and hotels rather than for factory operators.

The operator did not see the relevance of this to his job and since his understanding of English was not perfect, he "switched off" and did not take in and learn the messages. The operator had not intentionally ignored the training or forgotten it, because he had never understood the relevance of the training to his job in the first place. This example shows us how important it is to do the right type of training and to ensure that the reasons for doing it are understood.

2.1 SHORT- AND LONG-TERM MEMORY

We must remember that our best recollections are usually recalled as a result of some specific trigger. It could be a smell, a noise or some particular image that causes us to have quite vivid recall of a situation or information. The common factor about these events that we recall vividly from the past is that they were in some way memorable at the time. In order to be memorable, they will have had some significant impact on us at the time they happened. Unfortunately, sometimes this impact can be negative, in that something traumatic or harmful happened to us. On the other hand, we also remember things that have a very positive effect such as our wedding day or the birth of a child. Memory is perhaps the most fundamental factor in how human beings learn. It should inform all aspects of learning but often we don't consider it when planning our training. Think of the memory as a large expandable filing cabinet. Everything that we learn throughout our life goes into the cabinet. The cabinet doesn't actually leak or destroy information. The issue is one of information retrieval. We need to have a mechanism for getting that information out again. The challenge with learning is not so much about how to retain information, but much more about how to retrieve information that is already in our subconscious. Our brain is divided into two sides and most people have one side that dominates. If you are mainly analytical, logical and methodical, you are likely to be dominated by the left side of your brain, whereas if you are artistic or creative then you are more likely to be dominated by the right side of your brain. Neurologists tell us that our brain is made up of around 100 billion brain cells, which are called neurons. The information retrieval routes in our brain are known as neural pathways. With lack of use, some neural pathways temporarily close down in favour of others. We will be aware that knowledge that we use frequently is easily recalled and is the basis for our ability to perform routine tasks without perpetually needing to refer to an instruction manual. The more often that we recall information, the easier it becomes. This is known as reconsolidation. This is the basis for training in responding to high risk but infrequent situations – knowledge recall in organisations such as the Fire and Emergency Response Services is ensured by regular practice and the use of simulations. Our training processes need to be designed to ensure that the appropriate neural pathways are opened up to ensure that we get rapid and effective recall of relevant information. The less that we use some information, then the more difficult it is to remember or retrieve. This is even the case for some information that we once knew very well. I have owned many cars over a long working life. During each period of owning a particular car, I had no trouble remembering the registration number of that vehicle. This is because that was information that I knew I might need – it's always a bit awkward if the police stop you and you can't tell them the number plate of the car that you are driving! This is a piece of information that you are highly motivated to remember. However, no sooner do you buy a new car than your need to remember the old registration number is overtaken by a much more urgent need to remember the new one! The interesting thing is that very quickly, the old number which you once knew so well, gets forgotten. This is one of the major problems with learning – we are human and we forget. When I am chastised by my long-suffering wife for having overlooked something, my response is usually "I'm sorry my dear, but I forgot!" That may well have been the case, but it doesn't usually get me off the hook!

Our training needs to be designed to ensure that people remember the key messages or information that we are trying to convey. I once remember being told by a university lecturer that undergraduate teaching involved transferring the information from the lecturer's blackboard into the notebook of the students, without it entering the heads of either!! In these circumstances it is hardly surprising that students and teachers get bored. To be able to train people effectively is a gift, but it is a gift that can be acquired. It is always good to know when your training is appreciated. Some years ago, a trainee came to see me after a particular training session and said to me "You enjoyed doing that, didn't you!". The odd thing was that I had never really thought about actually enjoying it. On reflection he was absolutely right, and actually it is difficult to succeed as an effective tutor if you do not enjoy what you are doing.

As I mentioned before, I find that the most common failing of university Professors is that they are so knowledgeable about their own subject that they cannot understand how their students cannot immediately grasp Einstein's Theory of Relativity. Good training is not the same thing as intelligence or academic achievement. Some trainers seem to think that they have to instantly transfer all the knowledge that they have assimilated over the last 20 years and get the trainee up to the same knowledge level after half a day of training. The result is often one of overload. As soon as a trainee loses their understanding, they also lose their interest. For this reason, the trainer needs to identify in advance what are the key messages that he wants the trainee to remember. Teaching is much more effective in small bursts which are then periodically repeated and the content progressively expanded. It is very much the "drip feed" principle.

Key points not only need to be repeated, but they need to be made memorable. If we start off with 100% of new information in our memory, then an hour later we will only remember 44% and after a month it will be down to 20%. This is known as the Forgetting Curve (Fig 2.1) and was developed by the German psychologist Hermann Ebbinghaus in the late 1800s. It works on the principle of "use it or lose it".

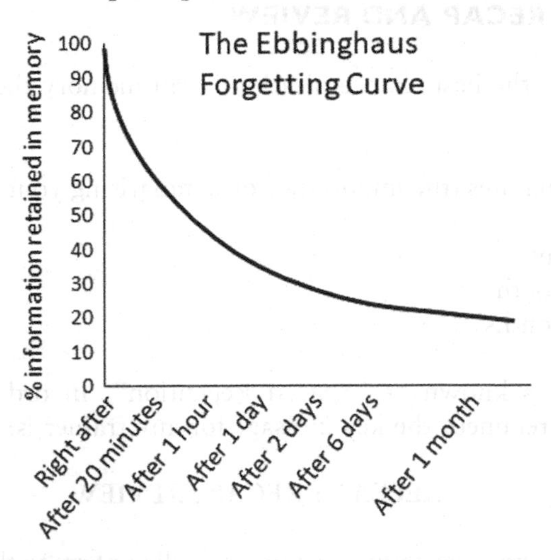

Figure 2.1 The Forgetting Curve.

Regular repetition hugely improves our ability to retain information. The repetition should not be left too long. The importance of summarising and checking understanding initially during the first training is crucial. In order to memorise things quickly, the first repetition should come within 15–20 minutes. If the training has a break for coffee or other personal needs then upon return, I would always remind the trainees of the key points that we have covered up to that point. I would also remind them of those same key points again at the end of the day.

In order to re-affirm the message, it is recommended that another reminder is given again after 24 hours. If the training is over several days, asking one of the trainees to summarise the previous day's learning at the beginning of each new day is a good way of getting involvement and recall and hence developing memory.

Initially, information is stored in our short-term memory. However, as the name suggests this is a rapidly decaying store. Most psychologists believe that short-term memory does not involve permanent changes to the brain and therefore memories fade quickly. One article that I read suggested that the short-term memory cannot handle more than five to seven items at a time. The long-term memory however appears to be permanent. A permanent change occurs in our brain when we remember things for a long time. Whereas the short-term memory can get overloaded, the capacity of the long-term memory is infinite. Therefore, we need to try and ensure that important training messages are transferred to our long-term memory. In order to retrieve information from our long-term memory it is often helpful to use a trigger or cue and we discuss this in much greater depth later in this chapter.

2.2 REPEAT, RECAP AND REVIEW

In order to have the best chance of a long-term memory then repetition is advised:

- After 20 minutes (the importance of summarising your message at the time)
- After 1 day
- After 1 month
- After 3 months

This technique is known as "Spaced Repetition". In order for training messages to be retained, the key message for any trainer is:

REPEAT : RECAP : REVIEW

The fact that you keep repeating the message will confirm to the trainees that you believe that this is an important point and so they are more inclined to remember it. In this book, certain points are repeated more than once in order for the reader to more easily memorise them. In some training circumstances, it may not be possible for you personally to carry out the repetition, but this is where you can invoke the help of the trainees' supervisors or mentors to do the reminders on your behalf. This is very powerful in training terms, because it introduces another person into the process who is reinforcing what you have said at the training. Not only that, but it also acts as a form of reminder/training for the supervisor or mentor.

Research carried out in the USA some years ago identified that if a manager issued a memo to his team requiring them to do something new and then he or she did nothing else to reinforce that this new practice was important, then the research showed that within six months the required practice would be being ignored. In other words, employees will tend to respond to what their boss sees as important. If you keep re-emphasising something and indicate that it is important to you, then they will remember that and do it. If you pay "lip service" to it on one occasion and then never check up or follow through, then people will think that it doesn't matter to you. If it doesn't matter to you then you can be sure it doesn't matter to them!

People will also remember things better if they themselves believe that what you are telling them is important. This is particularly significant if you are trying to convey some complex idea or skill. If people do not understand what you are saying, then they will not be able to remember it. This is particularly the case with health and safety training. Nearly everyone believes that they are safe – they believe that it's all the others who are not! Therefore, there is a tendency for people to come to safety training in the belief that they already know it all. It is important to make things as memorable as possible. When teaching children, we try and provide ways to make things memorable. When I am resetting the date on my old analogue wristwatch at the end of the month, I still repeat to myself:

"30 days hath September,
April June & November
All the rest have 31
Except for February alone
which has 28 days clear
And 29 each Leap Year"

I must have learned that at Primary School and despite my failing memory, I can still remember it! I am not suggesting that all your work instructions are converted to rhyme, but the trainer does need to think about how to ensure things are memorable. Things that you understand are memorised nine times faster. Remember also that things that are similar can get confused in the brain.

One of the problems of modern living is that much information is available from our computers and internet searches. This is hugely beneficial to us all, but a corollary of this is that we rely less and less on our memories because we rely so much on search engines. This means that we get less skilled at memory retrieval. It is therefore necessary to design our training in order to help people remember it. The following is a list of my top 20 tips on how to get people to remember your training:

The Top 20 – How to get people to memorise quickly:

1. Make it "fun".
2. Get them to take an interest in what you are saying.
3. If trainees are not interested then they will learn nothing.
4. Identify their needs at the outset and relate what you say to meet those needs.
5. Ensure that you prepare properly and don't spend time on teaching stuff that is irrelevant.
6. Use acronyms and mnemonics to help memorise.
7. Acronyms and mnemonics act as "cues" which can help trainees access information from their long-term memories.
8. Associate new words with those they already know.
9. Linking new learning to things that they already know increases its relevance and acts as a memory trigger.
10. Identify and highlight the most important information.

11. Even with the best trainer, people will never remember everything. Identify the most important bits and present those with high impact.
12. Structure the training so that it develops logically. Use flow diagrams rather than lots of text.
13. Ensure that your training evolves in a logical (stepwise) manner. The more logical it seems to the trainees, the more likely it is to be remembered.
14. Create a visual image. Memory is primarily "visual".
15. People remember visual images better than what they hear. We can often assimilate a pictorial message more easily than a convoluted verbal explanation.
16. Tell a story that illustrates the message.
17. People can relate to being told stories – provided that they are relevant. A trainer who is also a raconteur will retain the attention of the audience and demonstrate his or her knowledge and experience.
18. People remember more if they both hear and see information.
19. Remember to not just "tell people how it should be done", but provide practical examples using real equipment, props or projected slides.
20. Use practical experience – learn by doing.

Remember the old saying of the Chinese philosopher Xunzi (Xun Kuang) from the third century BC (and later claimed by Benjamin Franklin!).

"Tell me and I forget,

teach me and I remember,

involve me and I learn".

If you want to test people's ability to remember at the beginning of a training event try using the Word Recall exercise. This is an evolution of Rudyard Kipling's "Kim's Game". It is best done at the beginning of the training. The example given here (Fig 2.2) shows that you should use words that are in some way related to the training that is being carried out. This helps later in the training as it reinforces the use and importance of these words, even if the trainees are not entirely familiar with the words when they start the Word Recall exercise. However, using all 20 as new and unfamiliar words is likely to be de-motivating for the trainees.

A key bit of pre-training research that the tutor must do is to establish whether there are any people who cannot write, or who have a poor grasp of spoken English. This exercise is only suitable for participants who can write and read the language.

As can be seen, one copy of the Word Recall sheet is given to each participant. They are asked to study the words for three minutes. The tutor then asks them to turn the page over and write as many as they can remember on the blank side of the sheet. If the final scores are very low, this

WORD Recall

Study this list of 20 words for 3 minutes. When told by the tutor, turn over the page and write down as many of the words that you can remember:

Permit	Accident
Guard	Gloves
Safety	Exposure
Risk	Hard Hat
Hearing	Procedures
Training	People
Injured	Protection
Awareness	Respirator
Bandage	Earmuffs
Electricity	Hazard

Score one point for each correct answer.

Untrained = 4+

Improver = 8+

Master = 18+

Figure 2.2 Word Recall exercise.

will tell the tutor of the need to do even more REPEAT, RECAP and REVIEW during the session. If the participants do very well, it is a good idea to return to it later in the day and see how much they have remembered, as most of this exercise will have gone into their short-term memory and most of it will have been forgotten after a couple of hours.

A variation on this exercise for more practical training is to use a series of pictures, instead of the words as the subjects that have to be committed to memory.

When giving feedback on the exercise, remember that it is not your job to try and embarrass the participants. You should not go around and ask each person for their score. Just ask "*Have we got any Memory Masters*" and "*How many improvers do we have?*" Do not ask those with low scores to identify themselves.

One final point before we move on from the importance of memory to learning. The eminent psychologist, Professor Daniel T Willingham reminds us that we constantly overestimate how well we know something. He claims that both children and adults consistently "think their learning is more complete than it really is". This is an important message not only for the participants, but also for the tutor. It will undermine your credibility if you repeatedly do not know an answer. One trick to dealing with unanswered questions is to have a separate flip chart sheet that you designate to record questions that you do not know or cannot remember the answer to. This will quieten your critics at the time, but don't forget, if you commit to finding an answer later on, you must go back to that person (and to the other participants) with an answer once the training is over!

In our earlier discussion about the brain, we learned that it has two sides which have rather different functions. In the Top 20 list we also see that point 15 reminds us that people remember more when they both hear and see information at the same time. One of the reasons for this is that the function of speech comes from the left-hand side of the brain and images, colour, expressions and comprehension are processed on the right-hand side of the brain. Using both sides of the brain at once enhances its capability and the result is greater than the sum of the parts. Consequently, this means that using both senses of hearing and sight concurrently results in better understanding and better memory. People only tend to remember 10% of what they hear and 20% of what they read. This is why giving production operators a copy of the operating instruction to read once a year is a very poor method of training. It says more about the management attitude that they just want a signature once a year to somehow prove that they have done their job, rather than caring that their operators are trained and competent. However, if trainees both hear and see information at the same time, they can remember up to 80% of it. This is why training should involve both verbal and visual information – and if possible being able to put that learning into practice in a safe environment.

2.3 OTHER MEMORY TECHNIQUES – MNEMONICS AND ACRONYMS

A mnemonic is derived from the Greek word "mnemonikos", which means "**relating to memory**", and is the study and development of systems for improving and assisting the memory. It can be a word, phrase, saying or rhyme, whereas an acronym is an abbreviation formed from the initial letters of other words and pronounced as a word (e.g. RAF or NASA). An acronym is one form of mnemonic which can be useful in training and learning situations.

Modern life is full of acronyms, some of which we remember and some of which we do not. Often, we use and understand an acronym without understanding what the individual letters mean. For example, most people would know what a laser is, but few people could tell you that the word is an acronym of Light Amplification through the Stimulated Emission of Radiation. Such words as laser have now entered vocabulary as words in their own right and are not just acronyms. For an acronym to be a useful training aid it should follow the 4"R"s.

- Retain - Is the message simple?
- Recall - Is the acronym itself memorable?
- Relate - Does it form a word or group that trainees already know?
- Repeat - Can they repeat it?

In order for the acronym to be memorable, it must be simple and relatively short. Ideally it should form a recognisable word or sequence. That word or sequence might be unique to that group of trainees. I was training some safety, health and environmental auditors who worked for the multi-national company ABB. I wanted to get them to remember to observe all aspects of the sections of plant that they were auditing. The point was that the auditors could get a misunderstanding of how equipment complied with their standards by just looking at it from one side, whereas I wanted to encourage them to take a holistic view. I used the acronym ABB to remind them to look "Above", "Behind" and "Beyond" any one piece of equipment to get a complete impression. This acronym worked well for the ABB trainees, but would have been meaningless to people from other companies. The key point is that the acronym must be meaningful to the particular trainees and be significantly easier to remember than the message itself. If people can't remember the acronym, they have no hope of remembering the message!

We have just spoken about the importance of using both senses of speech and hearing when training. This is known as using multi-sensory messages. A useful memory jogger for the tutor is to use the acronym VHF, when preparing a new training package. Most people will have heard of the radio term VHF, which in daily life stands for Very High Frequency. Most of us may not be able to give a scientific explanation of VHF, but it is a recognisable term which is in common use. For the tutor we use the acronym VHF to mean:

V = Visual senses (Projection slides, Videos, Photos, Posters, Models, etc.)
H = Hearing senses (Tutor Speech, Participant Speech, Music, Sounds, etc.)
F = Feeling senses (Excitement, Boredom, Smells, Hunger, Thirst, etc.)

When developing a new training package, always remember to put yourself in the place of the trainee and ask VHF. Do not think only about the training itself, but also about the training environment. If you are training in a noisy workshop, will the hearing category be met and will people be able to hear what you are saying?

Another acronym that is in wide use among trainers is KISS. This means Keep It Simple Stupid! This is probably the most important message for tutors to remember when de-mystifying our subject. We often tend to overcomplicate our training explanations. In health and safety and particularly in process safety, there are some quite complex concepts and we need to break these down into easily understandable steps. More details of how to do this will be found in the Resources section of this book.

It is also quite possible to use the same letter repeated as an acronym. For example, the 5 P's of emergency management means Preparation and Planning Prevents Poor Performance or the 3 P's of safe working are People, Plant and Procedures.

The key point for trainers designing their own acronyms is that to be memorable, the acronym should be:

- Short (between 3 and 6 letters)
- Have impact
- Help with not only memory but also understanding
- Stand out – if you have too many acronyms people get confused

It can often be helpful to design your acronym around existing words or formulae. For example, people will readily remember things like:

- H_2O
- R2D2 (of Star Wars fame)
- BBC

What is even better is to relate the acronym to something that the delegates are familiar with in their daily work. In the case of the operation of a fork-lift truck, stability is all-important. So, an acronym for the daily checks that need to be remembered by the operator might be STABLE:

S – Steering

T – Tyres

A – Access

B – Brakes

L – Lights

E – Electric Charge

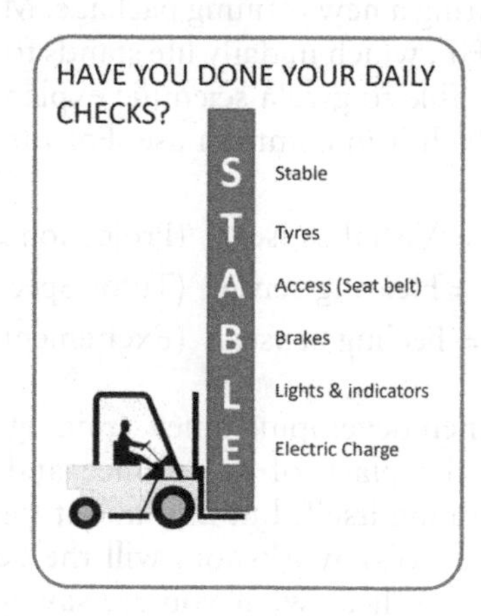

HAVE YOU DONE YOUR DAILY CHECKS?
S
T
A
B
L
E
Stable
Tyres
Access (Seat belt)
Brakes
Lights & indicators
Electric Charge

The advantage of using this type of acronym is that it acts as a cue to our memory and helps strengthen the neural pathways and allows us to retrieve the memorised information. It also means that if we compliment the initial training by putting posters around the workplace or in this case on the fork-lift truck itself, then every time the operator sees the poster/notice, their memory will be reinforced and they will be reminded of what they need to do.

It must be remembered that effective training is not just the initial training session, but it must be complimented periodically with reminders and refreshers such as posters, notices, supervision and audits. In fact, we should take every opportunity to reinforce safety messages.

When people are performing complex but routine tasks, such as airline pilots preparing for take-off, then a verbal cross-check is a very valuable memory jogger. The airline pilot uses a pre-prepared checklist to perform each flight preparation task. Each checkpoint is read out aloud by one pilot, while the co-pilot performs the action itself, pronouncing "Check" when the task has been completed or status confirmed. In addition to being a memory aid, this spoken checking system also acts as an immediate supervisory or audit check on that action, because the pilot not carrying out the practical action will be observing what is happening and should be able to recognise any mistakes or omissions. Although this approach is not commonly applied throughout the western world in industrial situations, I did notice that it was in common use when I was visiting manufacturing sites in Japan.

Repetition as an aid to memory retention requires considerable effort by both the tutor and the trainee. Initially the reinforcement should come quite quickly, say after 10 minutes or so. This is why it is so important to summarise the key points at the end of each section of training and to "tell them what you told them". Then as time passes the gaps between repetitions can increase. If the training is spread over a number of separate days or sessions, it is always a good idea at the start of each subsequent session to briefly ask or remind the trainees about what they learned up to the end of the previous session. Giving trainees a specific task to do in time for the next training session is also a way of getting the trainee to focus on what he or she has learned and to recall some information from their memory. The important message here is to remember that the more that the trainee is asked to recall from their memory about what they have been studying, the stronger those neural pathways will become and the more learning will be retained. This is why it is so important that training is done just before it needs to be practically applied, because workplace experience will immediately reinforce the trainee's memory.

2.4 GET IN THE CUE

Neuroscientist Professor Daniel Willingham of the University of Virginia calls memories "residues of thought". But simply thinking about something is not necessarily enough to create a memory. Why do we remember the things we do? You might remember a hotel that you visited during a past holiday, but not what you did while you were there. Why would you remember one aspect of that holiday but not the other? Much of what we remember is not a result of conscious effort. This is why some of the details of training sessions are so important. I often challenge my clients about their insistence that all their training courses use their house style for presentational slides, or that they always use the same training room. It can make things all too similar. If you want to have an impact so that people remember what they learned, then each learning event needs to be memorable and distinctive in some way. It might be the location, the surroundings, the exercises and activities or the presentations or even the tutor's charisma that triggers a memory in the trainee at a later date. Even if they don't remember all the detail, just pausing before doing the wrong thing and then going and asking for a refresher or some advice may be enough to prevent someone coming to harm.

Cues or prompts are what help us retrieve memories. Those who indulge in amateur dramatics will know the value of the prompter who stands in the wings of the stage. When an actor "drys" and momentarily forgets his words, the prompter does not need to reel off long sections of script. Usually, a short prompt refreshes the actor's memory and they do not require further help. The same thing can happen at work. The most common cause of this momentary loss of memory is a lapse of attention. There are many ways of creating memory cues in the workplace. We can rely on close supervision, checklists, recording of instrument readings, the use of "dead man's handles", audible alarms, signs, notices and auditing. The thing to remember is that your trainees are human, and therefore from time to time they will forget things. If the consequence of someone forgetting part of their training results in a critical event, then the training needs to be complimented by the provision of other cues or fail-safe measures.

2.5 MEMORY TECHNIQUES

Before you rush off to study at spaced intervals, creating clever mnemonic cues to help you with learning your part in the local amateur dramatic production, one final word of warning:

> We constantly overestimate how well we know something. Feeling that we know something is not a very accurate or reliable guide. According to Daniel Willingham we consistently "think we know more than we do" It is the concept of often "not knowing what we don't know!" A good example is doing DIY work at home. Lack of experience or

> knowledge has rarely stopped anyone "having a go" at that home-improvement task for which they are totally untrained. One day, a neighbour of mine rushed around to see me in a bit of a panic. He demanded to know if I had an Acrow prop? An acrow prop is a large piece of adjustable scaffolding, and not really in the average domestic tool kit. Unfortunately, I didn't have one. When I asked why, he took me to his home and showed me that he had started to remove an old chimney breast. Removing the old bricks with a sledge hammer had been easy enough, but he had overlooked the fact that the chimney breast was holding up a section of the floor above. He urgently needed to support the rapidly collapsing ceiling. He thought that he was competent to do the job, but he neither had the knowledge nor the right equipment! A key part of any training is knowing when to stop or not to start!

Willingham's "rule of thumb" is to study until you know the material – and then keep studying, for about another 20% of the time you've already spent. In other words, because we overestimate our knowledge, we should overlearn by about 20%. This is particularly important when learning to deal with situations which occur infrequently. The classic example is training to deal with emergencies. Thankfully, serious incidents are very rare, but if one were to occur, it is important to avoid the cognitive paralysis which displays itself as inaction in a crisis. This means that when dealing with a crisis the behaviours of key personnel must be instinctive. Even if they have never done it before, their training must automatically kick in and lead them to do the right thing. The way to achieve this instinctive behaviour is by repeated refresher training and practice. In this type of training, Willingham's rule of thumb could be expanded to "study until you know the material and then practice, practice and practice". The same concept applies to skills such as first aid, where serious incidents are rare, but regular practice is required to ensure that the emergency responders know instinctively what to do when the call comes.

The latest thoughts in training curriculum design from the United Kingdom, build on this idea of over-learning. The principle is to spend more time on fewer subjects, interleaving topics so that learners encounter them early on and then are exposed to them repeatedly over time, and using frequent low-stakes testing, spaced out over varying intervals, to stimulate deeper learning and recall.

In a world where it is so easy to use computer search engines to find information, the importance of human memory seems to be ever diminishing, and so trying to get trainees to develop better memory recall can seem counter-intuitive.

The human brain is an amazing device. It can allow us to do tasks that require a range of different inputs with apparent ease. If we take a simple task, such as making a cup of tea, then we barely even think about what we are doing. Our actions are almost automatic. In reality our brain is analysing information all the time and converting that into instructions for our bodies to perform. For example:

- Where is the kettle?
- How many people will want a cup of tea?
- Remove kettle lid
- Position kettle under the cold-water tap
- Turn on tap
- Assess water level in kettle
- Turn off tap
- Remove kettle to working surface
- Replace lid
- Plug into electrical socket
- Switch on electricity
- Find teapot
- Find tea caddy

  Etc. etc.

Our brain does all these things without us being aware that we are thinking about them.

Our family enjoy playing games of cards. In fact, my seven-year-old grandson can play card games quite easily if I explain the rules to him. Recently, I wanted to learn some new games, and bought a book of card games. The book is unbelievably complicated, even though when you eventually understand the rules, they are actually quite easy to play. It is actually surprisingly complex to explain how to do even simple tasks in writing. The important message for the health and safety tutor from this is that it is much more effective to show someone how to do something, rather than relying on telling them or using the written word. Training by "doing" is undoubtedly one of the best methods of training shop floor personnel in their work tasks.

One word of caution. One of the most common causes of accidents is when untrained people carry out potentially hazardous tasks. By getting a trainee involved in carrying out that task as a part of his or her training increases the chances of an injury occurring during training. In recent years, there have been more injuries among the military recruits during live-fire training exercises than have actually been injured in combat. The tutor must carry out a simple risk assessment to identify what might go wrong during the "hands-on" training, and ensure that controls are put in place to ensure that any errors during training will not lead to adverse consequences for the trainee or anyone else who could be affected.

"Memory" Checklist

New knowledge is initially held in the short-term memory and will degrade quickly. To increase retention time, the tutor should:

1. Make it "fun".
2. Ensure that the information is taught in an interesting way.
3. Get them to take an interest in what you are saying.
   - If trainees are not interested then they will learn nothing.
   - Ensure that the information is relevant to the trainee.
   - Ensure that the information can be understood (keep it simple).
4. Identify their needs at the outset and relate what is taught to meet those needs.
5. Ensure that you prepare properly and don't spend time on teaching stuff that is irrelevant.
6. Ensure the training is carried out just before it is needed in practice.
7. Use memory prompts such as mnemonics and acronyms as an aid to memory retention. These should be:
   - Short (between 3 and 6 letters)
   - Have impact
   - Help with not only memory but also understanding
   - Stand out – if you have too many acronyms people get confused
8. Regularly check understanding.
9. Repeat/Recap/Review key information.
10. Carry out the first repeat after 20 minutes.
11. Carry out the next repeat at the end of each day.
12. Invite trainees to recap the important messages of the training.
13. Involve supervisors and mentors in further longer-term reviews to reinforce the learning message.
14. Use signs and notices as memory joggers.
15. Associate new words with things and terms that they already know.
16. Link new learning to things that they already know increases its relevance and acts as a memory trigger.
17. Identify and highlight the most important information.
18. Even with the best trainer, people will never remember everything. Identify the most important bits and present those with high impact.
19. Structure the training so that it develops logically. Use flow diagrams rather than lots of text.
20. Ensure that your training evolves in a logical (stepwise) manner. The more logical it seems to the trainees, the more likely it is to be remembered.

21. Create a visual image. Memory is primarily "visual".
22. People remember visual images better than what they hear. We can often assimilate a pictorial message more easily than a convoluted verbal explanation.
23. Tell a story that illustrates the message.
24. People can relate to being told stories – provided that they are relevant. A trainer who is also a raconteur will retain the attention of the audience and demonstrate his or her knowledge and experience.
25. People remember more if they both hear and see information.
26. Remember to not just "tell people how it should be done", but provide practical examples using real equipment, props or projected slides.
27. Use practical experience – learn by doing.
28. Ensure that managers set a good example of what is expected.

Chapter 3

Communication – Getting the message across

Training is all about knowledge acquisition. Our memory is used to retain information, but that is not the primary source of new information for trainees at work. Particularly when we are younger, we rely on others' knowledge and advice. Education and training arise when that knowledge is communicated between two or more individuals. Good communication is an essential part of life and it is a key factor when attempting to influence the attitude, knowledge and skill of trainees during training.

Communication is not just limited to the imparting or exchanging of information orally. It includes all the senses, particularly hearing and sight. Many people tell me that they have been speaking, listening and observing things all their waking lives, but the problem is that despite all the practice we have had, a lot of people are just not very good at it, either that or they do not appreciate its importance.

Some managers that I have met over many years still think that communication is all about making pronouncements. They see it as "telling" people what is happening and what they need to do. They are not particularly interested in seeking feedback, as that might complicate things and might even suggest that they had got things wrong. Nothing could be further from the truth. Effective communication is a two-way process that involves some speaking and a lot of listening.

In Italy in the late 1800s, there was an economist by the name of Vilfredo Federico Damaso Pareto. It is said that he noticed that 80% of the peas harvested in his garden came from 20% of the plants. He later discovered that 80% of the land in Italy was owned by just 20% of the population. When he then investigated industrial output, he again found that 80% of production typically came from just 20% of the companies. His generalised conclusion, which is known as the Pareto effect, was that 80% of results will come from just 20% of the action. This 80/20 rule has also been applied to communication.

In the case of "active" listening, the 80/20 rule says that, in any conversation, the discussion leader should spend 80% of the time listening and only 20% of the time talking. This is not entirely the case in training situations, but it does indicate just how important listening is. Any training

DOI: 10.1201/9781003342779-4

session at which the tutor just lectures or "tells" the trainees what to do is unlikely to be very memorable or effective. Remember that the good Lord gave each of us two ears but only one mouth – it seems that he had always intended us to listen more than speak.

Communication is a form of interaction between people. Good communication at work is about understanding instructions, acquiring new skills, making requests, asking questions and relaying information. The good trainee will understand just how important it is to ask questions to seek clarification and understanding rather than just listening. Tutors should be aware that the ability to ask questions is one of the current limitations of some "e-learning" programmes. Good communication skills are perhaps the most basic skills that you can possess as an employee, yet they remain one of the most sought-after by employers. It is important to recognise that we can communicate in a range of different ways. As tutors, we need to understand the unintentional impact that we might have on our trainees.

3.1 VERBAL COMMUNICATION

Health and safety can be a subject that is potentially quite boring. The tutor's approach must be to find ways of stimulating interest. This can only be done if we understand who we are talking to. In this world of global business, it is increasingly common to find that we are training people who do not all speak the same first language – so the first question in communications is to check whether people are understanding the words that are being used (there are suggestions in Chapter 9 on how to deal with this).

The problem may not just be one language, it is very easy for the tutor to start using jargon which some trainees may not understand. Manager and "white collar" training tend to use more words and speech than training at the "shop floor" level, where practical experience is vital. However, no training can be done without the use of some level of the spoken word. A common problem is that people are just not listening. This might be that the trainee is just not interested in the subject, or it may be that the tutor is not having a great enough impact. This situation can arise from one or more of a number of possibilities:

- The tutor may have said the wrong thing or
- The tutor said the right thing, but
  - The trainee did not understand, or
  - The trainee mis-heard what was said, or
  - The trainee mistook what was said for something else
  - The trainee was not paying attention for some reason

The objective for the tutor is not just to get through to the end of training session, but it is to ensure that the trainees continuously apply the new knowledge in the way that was intended and on every occasion.

The Austrian Nobel Prize winner Konrad Lorenz summarised this situation very well. He reminds us that:

"Said is not heard

Heard is not understood

Understood is not agreed upon

Agreed is not yet applied

Applied is not necessarily maintained"

It is clear that training involves a number of steps between the tutor conveying the first message and then that message being consistently applied. The tutor is in a position to facilitate effective communication by the way he or she plans and executes the training. The greater the engagement, involvement and enjoyment that the trainees have in the subject, the greater will be the likelihood that they will understand and re-apply the learning. In order to communicate effectively with the audience, the tutor will need to display the following characteristics:

Training Communication Characteristics (the C's of training communication)

- Courage — Dare to be different.
- Conviction — Be knowledgeable about the subject.
- Coach — Lead and mentor trainees – don't just "lecture or talk at".
- Care — Demonstrate that you are here to help.
- Consciousness — Be aware of the effect that you are having on the audience.
- Consideration — See the other trainee's point of view.
- Conciseness — Decide what are the key messages and get those across.
- Clarity — Get to the point. Do not digress unnecessarily.
- Creativity — Surprise them with your approach. Think out of the box.
- Correctness — Ensure that information that you communicate is correct.
- Complete — Complete all of the key facts in the allotted time.
- Concern — Be aware of the impact that you are having.
- Confidence — Speak authoritatively.
- Certainty — Be definite.
- Culture — Allow for cultural and language differences.
- Criticism — Respond to criticism in a calm and positive manner.
- Conduct — Be aware of the effect that your appearance and movements may have.

The ability to be persuasive is an important skill in making training communication have high impact. Tutors with strong persuasion skills will tend to have a keen awareness of their audience. They are usually friendly and charismatic, and can establish a rapport with trainees at several different levels. Persuasive tutors make their messages or ideas tempting, by presenting them in such a way that the trainees believe they will personally benefit from applying the training. The ability to be persuasive comes from being confident, knowing your subject and answering the question without appearing aggressive or condescending.

Do not worry about feeling nervous. Nervousness is a sign that you are wanting to do your best and even the most experienced speakers will be nervous before the event. If you are well prepared and knowledgeable, your nerves will disappear once you start to talk.

Communication is a combination of verbal, visual, tone of voice and body language signals. This is why it is so important for the tutor to convey enthusiasm and to vary the tone to create empathy with the trainees. Body language can convey both intended and unintended signals. Even small details such as different dress standards between the tutor and the trainees can create a barrier which makes information exchange more difficult.

Professor Albert Mehrabian of the University of California, Los Angeles (UCLA), is an authority on communication. When evaluating how we feel about someone else he surmised that 7% of those feelings arise from spoken communications, while 38% take place through tone and voice and the remaining 55% of communication of these factors take place through the body language we use (particularly our facial expressions). Clearly, we as tutors are being assessed as soon as the trainees walk in the door. If we can convey a feeling that we are competent and can speak in a way that the trainees will understand, then the trainees are much more likely to be engaged with the subject. What Mehrabian's work tells us, however, is that when it comes to how the trainees feel about being in the training session, it is less about what we actually say and is more about our body language and tone of voice. If we come over with a tone of voice that is school master-ish, then our audience will not respond well. Equally if we appear interested and enthusiastic through our energy and body language, then the trainees will respond well, in return. The key point in establishing a good rapport between the tutor and trainees is that we establish an atmosphere of mutual trust.

3.2 ACTIVE LISTENING

Winston Churchill once said that "Courage is what it takes to stand up and speak, Courage is also needed to sit down and listen". Listening is one of the most challenging skills for tutors. Mastering the technique of "active listening" will enable tutors to get their trainees integrated into and contributing to the training session. As training leaders, we sometimes tend to focus too much on delivering our message and too little on listening to feedback. Active listening involves paying close attention to what others are saying, and asking clarifying questions to show that we are both interested and understand the point that is being made. This facilitates effective communication while simultaneously allowing us to show respect and also build relationships with the other trainees. It is quite easy to know if this two-way process is working, because if the trainees feel that their input is not valued, they will cease to be engaged and will not respond. No response is still feedback to the tutor – and a very powerful and uncomfortable one!

To demonstrate active listening, we need to consider the following:

1. Pay attention to what is being said by the trainees. Turn and focus your attention on the person speaking. Make eye contact. Demonstrate understanding of the point being made by showing that you are listening by nodding, saying "Yes" or by facial expression.
2. How does the trainee feel? Are they demonstrating anger, boredom, sharing knowledge or have they just misunderstood what the tutor said?
3. Look for non-verbal signals that convey messages of active interest, confusion, anger or boredom and respond to those.
4. Remember to either answer the question before you move on, or take a note of the question and come back to answer it before the end of that training session.
5. If the trainee makes a supportive comment, thank them and repeat it in a way that reinforces points that you as the tutor have already made or are about to make.

The most important thing about communication is hearing what isn't being said – reading between the lines.

3.3 VISUAL COMMUNICATION

A picture paints a thousand words. One of the most memorable ways of communicating is to see something that has substantial impact upon you. We have all had experiences in life where the image sticks with us for a very long time. It may be witnessing a road traffic collision, or seeing a loved one in hospital with some terminal disease or seeing how our own forethought avoided some serious injury. We will have learned from these "high impact" experiences and typically the images will return to us periodically and will help us avoid us repeating the same mistake. I have had the responsibility of investigating fatal accidents over the last 20 years. For me, each one of those has been a life-changing event. The first one had the most impact upon me and involved a building collapse. At that time, I was involved in engineering management, but the significance of that event, where I was called to be involved in the incident enquiry team, had such a profound effect upon me that from that time I chose to focus my career on health and safety improvement. The same benefit can arise in a training environment.

Showing diagrams or photographs of situations that have led to injuries is a very powerful way of making a point. This visual communication can have even higher impact if the trainees themselves are asked to say what could have gone wrong and how could it have been avoided. Trainees can be reminded of the impact of these images by the use of posters and signs around the workplace.

However, the greatest impact in terms of communication comes from practical application. Benjamin Franklin is claimed to have said:

"Tell me and I forget

Teach me and I remember

Involve me and I learn".

Good training that has high impact, needs to ensure that the trainees are involved in what they are doing and not just attending. This is crucially important for practical workers at the shop floor level who are not used to sitting in meetings and offices. Normally they will respond much better to training that involves them. This could involve setting up demonstrations of equipment where they can "learn by doing", or by using mock scenarios (see Resources section "O"). Shop floor workers in particular will learn much more quickly by seeing and applying their new skills rather than just being talked at. Training will always be more effective if it involves "pulling or leading" the trainees by encouraging them to develop their existing skills rather than just "pushing or telling" them what to do all the time.

It is crucial when using practical experience during training sessions to make sure that accidents and injuries cannot happen. In these situations, carrying out a risk assessment of the training task when trainees are involved is essential.

Visual information in the form of flow diagrams is also a simple and effective way of conveying a sequence of actions where decisions may need to be made. Shop floor workers will appreciate this approach rather than having long-winded text descriptions contained in dry and dusty job instructions.

Many visual health and safety communications can be achieved without the need for a tutor or mentor being present. These will normally be in the form of posters, hazard warning signs and e-learning packages. It is important with visual messages that are in daily view that they do not get stale and are regularly replaced and updated. In training centres and training rooms it is important to get visual messages across to trainees at every opportunity. Remember that trainees may arrive early or be at a loose end during lunch breaks and so make sure that there are displays, posters or messages that they can browse through while waiting for a training session to start. Provide something relevant, interesting and concise and you will be surprised how many people will look at it.

3.4 WRITTEN COMMUNICATION

Written communication as a part of training will depend on the purpose of the writing. It could be for reference purposes after the training is completed, in which case the reader will have more time available for reading. However, if the written communication is to be read during the training session, it needs to very clear and concise.

1. Ensure that the writer understands the purpose of the writing. Brevity is more important for training handouts than for reference material.
2. Clearly state the purpose of the document.
3. Keep the language simple and appropriate to the group being trained.
4. Focus on the task.
5. Use diagrams, photographs and flowcharts to aid understanding.
6. Clearly state who needs to do what and when.
7. Define the outcome (when does the trainee know when to stop?)
8. Get a colleague to proofread and do a sanity check on the document.

Remember that good written communication involves being able to convey information to that particular trainee group, in a simple and unambiguous way. It could be that different levels in the organisation following the same training may require different written support documents.

Communication Checklist

Your message will only be remembered if there is effective two-way communication with the trainees:

1. Verbal, visual and written communication is important.

   "Said is not heard
   Heard is not understood
   Understood is not agreed upon
   Agreed is not yet applied
   Applied is not necessarily maintained"

2. Tutor's Training Communication Characteristics (the C's of training communication)

   - Courage — Dare to be different.
   - Conviction — Be knowledgeable about the subject.
   - Coach — Lead and mentor trainees – don't just "lecture or talk at".
   - Care — Demonstrate that you are here to help.

• Consciousness	Be aware of the effect that you are having on the audience.
• Consideration	See the other trainee's point of view.
• Conciseness	Decide what are the key messages and get those across.
• Clarity	Get to the point. Do not digress unnecessarily.
• Creativity	Surprise them with your approach. Think out of the box.
• Correctness	Ensure that information that you communicate is correct.
• Complete	Complete all of the key facts in the allotted time.
• Concern	Be aware of the impact that you are having.
• Confidence	Speak authoritatively.
• Certainty	Be definite.
• Culture	Allow for cultural and language differences.
• Criticism	Respond to criticism in a calm and positive manner.
• Conduct	Be aware of the effect that your appearance and movements may have.

Remember that the words that you use are only a part of the impact you create. Your tone of voice and body language will have a big impact on the audience.

3. Listen to what the trainees say, and respond. (Active Listening)
4. Visual communication – "A picture paints a thousand words"
5. Communicate by involvement:

   "Tell me and I forget
   Teach me and I remember
   Involve me and I learn".

6. Written communication
   - Ensure that the writer understands the purpose of the writing. Brevity is more important for training handouts than for reference material.
   - Clearly state the purpose of the document.
   - Keep the language simple and appropriate to the group being trained.
   - Focus on the task.
   - Use diagrams, photographs and flowcharts to aid understanding.
   - Clearly state who needs to do what & when.
   - Define the outcome (when does the trainee know when to stop?)
   - Get a colleague to proofread and do a sanity check on the document.
7. It is completely normal to be nervous at the start of your presentations.

Chapter 4

The audience – Who will be there?

4.1 TYPES OF TRAINEES

We have no control over the personalities of the people that we need to train. Some of these will be difficult to deal with but it is important to remember that you can't control how other people behave, you can only control the way that you respond to them.

It is therefore important for a tutor to understand that different personality types may need different approaches to training. To understand this, it is important to recognise that trainees will have different reasons for attending the training. These are described by Tom Stapleton* as the 3 "V's" – Voluntold (told to volunteer!), Vacationer and the Valued Learner. These three categories of trainees have very different reasons for attendance and therefore very different motivations and willingness to learn. For each of the three Stapleton categories, the tutor must be prepared to adapt his or her approach.

a. Voluntold: They have to be there, but they would rather be somewhere/anywhere else. Contributing little to and getting little from the training, they may have a chip on their shoulder and can be disruptive. It's best to coax – not confront this type.
b. Vacationer: They're happy just to be away from the job. Typically, indifferent about being there, they often get as much as they give. This type can be a challenge, but often responds favourably when asked, for example, to work in a team setting.
c. Valued Learner: They're positive about attending the training and participate interactively. And they usually speak up and contribute to enhancing the experience for all. This type often can be an inspiration to the other participants.

It should be remembered that it is even possible to demotivate the Valued Learners if the training delivery, material or relevance is not appropriate.

* Courtesy of Tom Stapleton: Stapleton Communications Glendale, California, USA.

DOI: 10.1201/9781003342779-5

The choice of the type of training and its design should take into account the demographic of the intended audience. The older members of staff are more likely to be comfortable with traditional teaching methods, involving presentations and exercises, whereas younger people are much more at home with using computer-based learning and I.T. in general. It is also important to decide at an early stage whether the training is to be individual or group-based. Specific skill training is often best done on a one-to-one or small group basis, whereas general knowledge transfer can often be done in larger groups.

Large groups can easily suffer from group dynamics where involvement of all attendees can be a challenge. The layout of the training room can often help minimise this problem (see section 4.3 of this chapter), but there will always be a mix of personalities in any group. The tutor needs to be aware of certain personality types and how he or she might be able to deal with any negative influences. The table below shows some of the personality types that may be involved in the training. The second and third columns show what the consequences might be and also how those can be controlled or minimised.

TABLE – Typical personality types at a training session

Personality type	*Consequence for the tutor*	*How to deal with it*
Eager	Often raises a lot of questions. Which might lead to time over-runs	Answers the questions Ask others what they think Suggest that we deal with it afterwards
Joker	Can cause diversions Could ridicule the tutor Often done as an aside in larger groups Might be a useful ice-breaker	Remember you can teach anyone anything if they are smiling! Ask them to share it with everyone. If it is destructive move on by saying "but seriously, what would you do?"
Voluntold (would rather be somewhere else)	Not engaged or interested	Involve them directly Help them understand why the subject is important
Sceptic	Negative about everything Destructive	Involve then directly Ask "Why" they have their view and present arguments to the contrary Defer questions until later in the training when hopefully things will be clearer
Busy	Arrives late resulting in the need to recap Continuously dealing with texts and emails Not committed to the training	Recap Ask attendees from the beginning as a courtesy to switch phone off or to silent Keep them occupied

(Continued)

Personality type	*Consequence for the tutor*	*How to deal with it*
Silent	Can go through the training without being involved Tutor unable to judge whether they are understanding the training	Engage them "What do you think" or is that OK? Talk to them on a one-to-one during breaks Ask them a question that checks their understanding
Fiddler	Has irritating habit that distracts you Usually a sign of loss of concentration	Take a break Speed up the presentation Move into an exercise or do something practical
Boss	Can dominate the topic Tends to put a dampener on things Others may just say what they think the boss wants to hear	Speak to him in advance and ask him to stay in the background Ask boss to do the introduction (i.e. show commitment) and then just observe.

Difficult trainees are not only the ones who are aggressive. The very quiet or introverted trainee can be equally difficult to handle. The key is to understand why they are behaving in the way that they are. The very quiet ones are often the ones who are failing to understand what is being said or demonstrated. It may even be that the tutor is speaking in a language that is not the trainee's first language, or is using technical terminology that is not understood, or just going too fast. In these circumstances the tutor needs to regularly check individual's level of understanding. Something as simple as "Is that OK Sue?" may be enough for the trainee to ask for more clarification. Difficult trainees may be exhibiting one or more of a range of different emotions such as aggression, frustration, ignorance, bravado, negativity, boredom, experience or shyness. The way in which the tutor deals with these is to follow the eight golden rules of response:

The Golden Rules for Trainers responding to argumentative or difficult trainees:

1. Self-control (do not overreact to arguments)
2. Listen
3. Seek further information and understanding through questions
4. Show interest (show that you have listened)
5. Explain why you said what you did
6. Recognise any valid point made
7. Show appreciation for their contribution
8. Follow through with any promise you make

The most important of the golden rules is that of "self-control" and the need to pause and think before we respond to an antagonistic comment.

Benjamin Franklin, one of the Founding Fathers of the United States of America, reminds us all that we should:

> *"Remember not only to say the right thing in the right place, but far more difficult still, to leave unsaid the wrong thing at the tempting moment."*

It is inevitable that the tutor will need to deal with difficult people from time to time, but there is nothing more rewarding than completing a training session that has been well received. I believe that training is a great learning event for the tutor and the more that we do, the better we become at it. As Dr Troy Amdal, the Co-Founder of The Oola Life Coach Network in the USA, reminds us:

> Be thankful for the difficult times.
> During those times, you grow.
> Be thankful for your limitations,
> Because they give you opportunities for improvement.
> Be thankful for each new challenge,
> Because it will build your strength & character.
> Be thankful for your mistakes.
> They will teach you valuable lessons
>
> *Troy Amdahl*

The tutor's personality can also have a significant effect on how trainees respond to the training. I was involved in an international health and safety training event in a country in Europe some years ago. I had finished leading my session and was followed by a tutor who I did not know particularly well. His session was explosive! He went straight into firing loaded questions directly at pale-faced individuals who invariably did not know the answer. It was not only humiliating for them, but it was downright scary. Even I, who was observing from the back of the room, hoped that he wasn't going to turn on me for an answer. I later learned that this tutor was a military reservist who had been trained in the art of interrogation of potential enemy spies! He taught me a valuable lesson about how not to win the hearts and minds of the trainees!

It can be helpful to establish some guidance for trainees on how to get the most out of any training event. There are nine key messages. These can be communicated verbally at the beginning of the event, but I prefer to have it somewhere that it is visible throughout the training. The best place for these is either on a wall-mounted poster at the front of the training area, or for training courses; either contained within the handouts manual or even better printed on the back of the place name toblerones. The messages are:

1. Enter into the discussions ENTHUSIASTICALLY
2. GIVE FREELY of your experience
3. CONFINE your discussion to the topic
4. Say what you THINK
5. Only ONE PERSON should talk at a time
6. LISTEN CAREFULLY to what is being said
7. BE PATIENT with other trainees
8. APPRECIATE the other trainees' points of view
9. BE PROMPT

I always like to start any training event with a request, saying "if anything is not clear or you don't understand what I am saying or I am going too fast or too slow, please stop me and tell me".

4.2 LARGE AUDIENCES

Teaching large audiences is a completely different skill to small group or one-to-one training. Not only that but it can be much more intimidating for the presenter! There are multiple reasons why this can be so challenging.

With a large audience, the presenter cannot be sure why each individual is there. Are they there to learn about the subject, or are they just Voluntolds who have been told to come? It is almost certain, that even if the delegates are all wearing name badges, you will not be able to see the badges and therefore cannot easily address them by name. Getting interaction going between the presenter and individuals in a large audience is difficult and even when it happens it can lead to time management problems and the session going into uncontrolled time over-run.

The presenter needs to come over as highly professional. Over-running your time is a complete taboo. Practice your input repeatedly to make sure you know how long it is going to take. The great danger for the expert at this type of event is "talking off the cuff". This will almost certainly go wrong, as it will inevitably overrun the allocated time. There are a few professional speakers who can make it appear as though their comments are "off the cuff" but in reality they are very well-rehearsed.

Whatever happens, do not read your speech – you may as well give them a book to read and it will be really difficult to create any real impact with the audience.

Remember the old adage of any presentation:

- Tell them what you are going to tell them
- Tell them
- Tell them what you told them

4.2.1 Tell Them What You're Going to Tell Them

What this means is to open your talk by briefly explaining what you are going to talk about. Tell them at what stage you will plan to take questions. I would suggest that you allocate some limited time at the end and then offer to continue the discussion after the talk is over or in the bar later. Ideally, have an independent chairperson or master of ceremonies to control the number and time allocated to questions, so that if time runs out, it is the chairperson's responsibility to manage the situation and not yours!

4.2.2 Tell Them

Then tell them the detail – this will be the major part of your talk. Remember to stay on message and be aware of the clock. A simple method that I have seen used is a narrow, coloured bar running down the side of your slides. When the bar is halfway down the slide you know that you should be halfway through your talk. When the bar is all the way down you have reached the end.

Alternatively, having a trusted colleague sitting in the front row, with a small prompt board indicating how your timing is going (e.g. "Speed Up" or "5 mins left") can be very helpful as even with a clock in front of you, it is very easy not to see it.

4.2.3 Tell Them What You Told Them

And then finally summarise the important points of the talk; these will be the messages that you want them to take away with them. (Remember the "Repeat" message from Chapter 2.)

Make sure that your presentation is suitable for the venue. Some auditorium projection systems do not handle portrait orientation slides well. Before the talk, check that the fonts and the colours that you use are visible at all places in the auditorium. Some projectors automatically resize your image to extra-wide screen leading to a distorted effect on photographs. The biggest danger is where you embed videos into the presentation. In my experience these have a huge potential to go wrong. Many years ago, I was asked to provide an input into a health and safety training day for a large group of Chemical Process Operators. My session was scheduled for part way through the day. I was travelling quite a distance to be there and so I did not have the opportunity to go into the venue before the day started to check my equipment. However, I had used that venue before without any trouble. My session hinged around a series of video clips that were embedded into my presentation. The night before, I had taken the precaution of checking that my presentation worked as required on my laptop, having been assured many times by the IT whizz kids that if you can see it on your laptop, then the same image will get transmitted through the LCD projector onto the screen. Except it doesn't always! I joined the training day during their mid-morning break and only had a few minutes to deal with the niceties and connect up my kit. Everything went fine with the presentation until "lo and behold" when the first video clip came due, it showed as planned on my laptop screen, but was not replicated to the main screen. The problem was that all the interaction that I had planned was based on discussion of the video clips – so no video clips = no discussion. It got worse, because there were multiple video clips in the presentation and none of them showed on the screen. If it had been a small group, I could have sent them off for an early coffee break while I sorted it out, but with this size of audience that was not an option. I had no opportunity to re-group and things just went from bad to worse. So the message is – if you are using materials that are critical to your message, make absolutely sure that everything works as you intended before you start.

4.2.4 Ergonomics

Large audiences can often arise in places like theatres where there is a custom to follow the traditions of watching plays and shows, where the audience is in the dark and the performer is illuminated on the stage. As any actor will tell you, they cannot see the audience further back than the first few rows of the stalls. This is a really difficult arrangement for a speaker or lecturer because it is essential to get feedback from your audience. If you cannot see most of the audience, then it is very difficult to get visual feedback and you won't really know what you got right and what you got wrong. It is also much easier for the hecklers to remain invisible! I would always recommend that the auditorium should be wholly or partially lit so that you can see whether you are having the desired impact or not.

The other problem with having the auditorium lights out is that it often forces the speaker to retreat behind a lectern. This has the benefit that he or she is less likely to stumble off the front of the stage and land in a dishevelled heap at the feet of the front row of the audience. The problem is that standing rigid behind a lectern is very formal and does nothing to engender excitement or engagement with the audience. I like to put lots of energy into my presentations and that involves moving around a lot. And you can only move around safely if the lights are on!

Think about whether you will expect the members of your audience to write anything down. For example, if you are providing any handouts or want them to write something, then the theatre venue is not appropriate as there are no tables or places to write, whereas a professional conference venue or a university lecture room will invariably have writing desks or facilities available for the audience. If your audience are not planning to write anything down, then they will remember less.

Before agreeing to present to a large audience, ensure that they will be able to hear you. If it is a large venue, you will almost certainly need a microphone. If like me you move around during the presentation then check that you will have a personal radio microphone. There is nothing worse than responding to a question in the audience only to find that your voice disappears because you turned your head away from the fixed microphone! Remember also that if you are anticipating responding to questions in large auditoria, then you must be able to hear the question. There is nothing so frustrating for a questioner than to have to repeat their question several times because the speaker doesn't hear it. In these situations, in addition to the speaker's radio microphone, it may also be necessary to have one or more "roving" microphones for the audience to use.

4.3 TRAINING ROOM LAYOUTS AND NUMBERS OF TRAINEES

Something as simple as the arrangement of the chairs and tables in the training room can make the difference between a successful and an unsuccessful training session.

The room layout needs to facilitate the tutor's ability to engage with the trainees and to minimise distractions. It is essential that the tutor can be seen easily by everyone and that any training aids that are used (such as VDUs, projection screens, models and displays) can be easily seen. The key point is that if the trainees are seated in rows, the ones at the back will not be engaged and can quite easily be distracted. The training room layout will be dependent on the room size, shape and the flexibility of the furniture. The desired layout of the training room may influence the choice of training location.

The first criterion in selecting the location is the number of trainees to be accommodated. If the training is a factual communication of a new safety policy or change in legislation where there may be limited need for discussion, then it may be appropriate to do that with a large group, who may be arranged in rows in what is known as a theatre-style layout. In most training courses then the maximum number of trainees should not be greater than about 15–18 people. The precise numbers should be calculated on the basis of what activities are planned. If syndicate working is planned then effective syndicates are typically five or six trainees. There will normally be either two or three syndicate groups so that sort of training session overall numbers will be between 10 (2 × 5) and 18 (3 × 6). The more detailed the practical work, the smaller the numbers that can be trained at any one go. I regularly run health and safety auditing training which involves a high percentage of time carrying out "mock" audits. Doing an audit involves walking around a workplace and doing this with a large group of trainees can be intimidating for the auditees and so that exercise is limited to two or three trainee auditors and a mentor. This means that for practical reasons these training sessions are limited to six trainees per session. Very often training of large numbers of people is arranged by deciding how many people we can squeeze into the room, whereas it should be based on what is the number needed to make the training activities effective. Having identified how many trainees can be trained at once, it is simple to decide how many sessions are needed to cover everyone who needs to be trained. There will be many occasions where the training is so practically based, that the number of trainees/tutor need to be one-to-one.

4.3.1 Board Room Layout

Once the number of trainees/event has been decided, it is then necessary to decide on the room layout. The "board room" style of meeting room (Fig 4.1) is rarely suitable for training. This type of room often has a single large table that cannot be re-positioned. The result is that the tutor is compelled to be at the head of the table with most of the trainees looking at each other across the table, rather than looking towards the tutor. Board room style layout is often used for training that requires all trainees to have access to computers and can also be useful for syndicate style break-our exercises, but for most group training the board room layout has significant drawbacks.

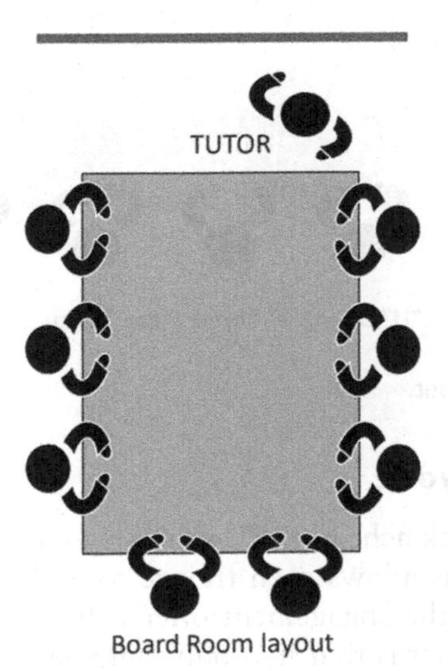

Figure 4.1 Board room style layout.

4.3.2 U-Shape or Horseshoe Layout

Setting the tables and chairs in a "U" shape is my preferred layout. This allows the tutor to move around and to approach individuals so that they can be addressed personally. This is a huge benefit in terms of getting them engaged. It also allows for smaller groups working without the need to break out into different rooms. The U shape (Fig 4.2) has three sides and so it is quite easy to divide the trainees into three groups to do quick collaborative work without losing the time to break out into separate rooms. This layout is also good when it comes to demonstrating a practical task such as the use of a noise meter, because by standing in the centre of the horseshoe, everyone is close enough to see. To get the maximum number of trainees looking directly forwards at the presentation screen, the back of the U-shape should be longer than the two sides. The drawback of this arrangement is that like in the boardroom, there are some people sitting at right angles to the screen who may develop neck ache. This layout also requires quite a large room to be effective.

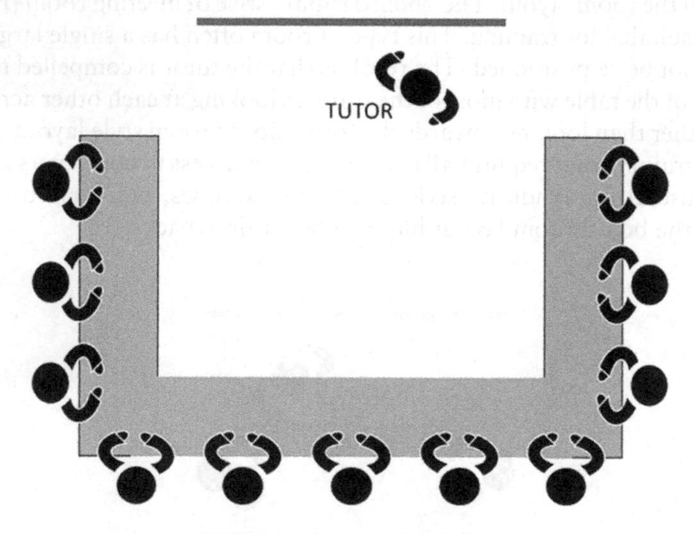

Figure 4.2 "U" shaped layout.

4.3.3 V-Shape Layout

The solution to the neck ache in the U-shape layout is to move to a shallow V-shape (Fig 4.3). This allows each trainee to be looking forwards at the presenter and retains the engagement offered by the U-shape layout. The drawback for this layout is that it is only suitable for small groups – typically a maximum of six.

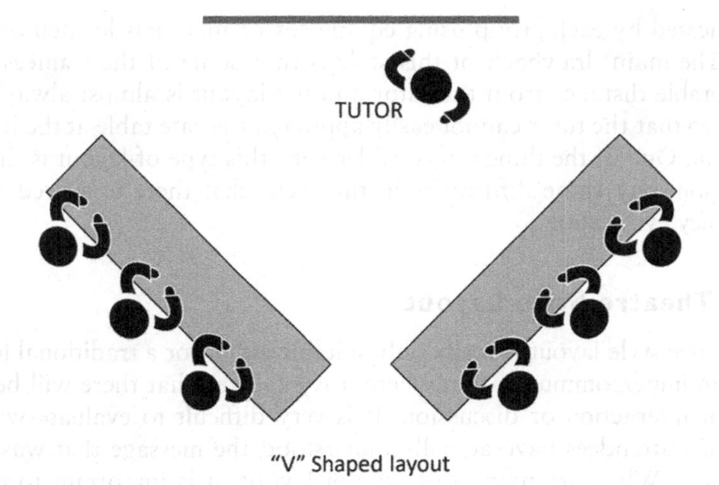

Figure 4.3 "V" shaped layout.

4.3.4 Cafe Style Layout

The Cafe Style layout is suitable for large groups where there is a need to work as teams. These might be natural work groups (i.e. a shift team), teams of the same first language speaking or just ad hoc groupings. Typically, this layout (Fig 4.4) requires a very large room and ideally requires round tables.

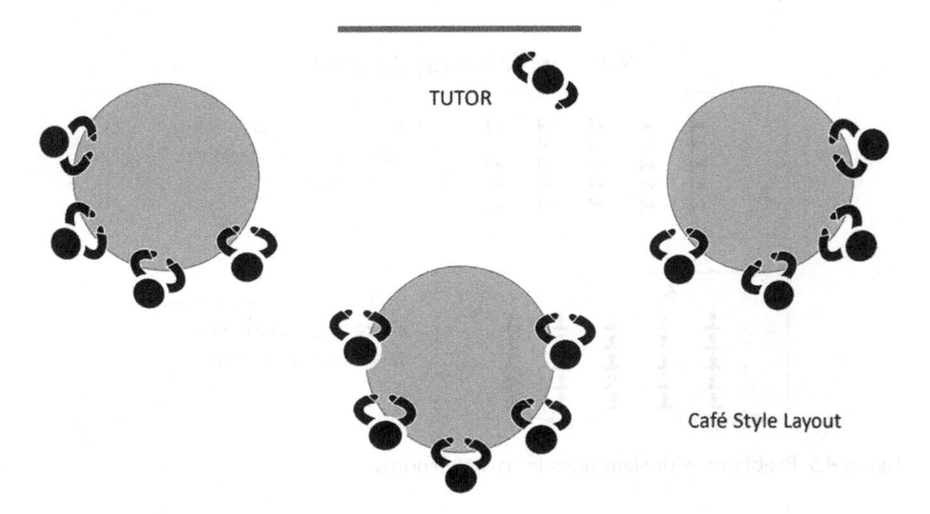

Figure 4.4 Café style layout.

Care has to be taken to ensure that some of the trainees are not left with their backs wholly or partially to the main speaker. This layout can be advantageous if the tutor is doing a practical demonstration that needs to be replicated by each group using equipment or materials located on their table. The main drawback of this style is that some of the trainees are a considerable distance from the tutor and the layout is almost always congested, so that the tutor cannot easily approach the cafe table at the back of the room. One of the things to consider with this type of layout is that the congestion may cause difficulties in the event that there is a need for an emergency evacuation.

4.3.5 Theatre Style Layout

The theatre style layout is really only suitable either for a traditional lecture or for training/communication where it is expected that there will be little need for interaction or discussion. It is very difficult to evaluate whether individual attendees have actually understood the message that was communicated. When arranging this type of layout, it is important to ensure that emergency evacuation regulations are complied with. In some countries this means that seats may need to be linked together to prevent them becoming trip hazards in the event that there is a stampede for the exit.

Seating layout in rooms that are narrower than they are long can cause problems, if the chairs are laid out the wrong way. Laying chairs out so that some people are a long way from the screen will lead to problems for the tutor in keeping attention and getting interaction. Likewise, having trainees sitting at a very acute angle to the projection screen will also make understanding of the message difficult (Fig 4.5).

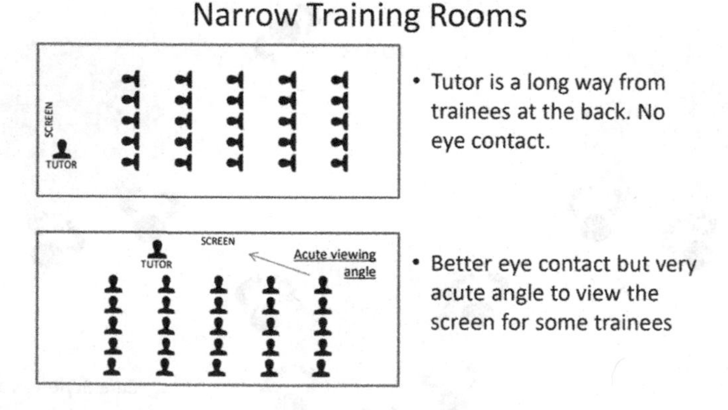

Narrow Training Rooms
SCREEN
TUTOR
• Tutor is a long way from trainees at the back. No eye contact.
SCREEN
TUTOR
Acute viewing angle
• Better eye contact but very acute angle to view the screen for some trainees

Figure 4.5 Problems with sight lines in narrow rooms.

One final point in relation to seating. Put yourself in the position of the trainees. They may be needing to sit down for long periods of time. It is a good idea to ensure that the seats are sufficiently comfortable to allow the trainees to complete the training session without getting a stiff back or numbing to other parts of their anatomy!

4.4 WHERE SHOULD TRAINING BE DONE?

Before deciding on the format and location of the training, it is necessary to be clear about to whom the training is targeted. The type of training that is appropriate to members of the Board of Directors is likely to be very different to that required for process operators required to operate a high-tech distributed control system. There is a difference between training and education. Both are different facets of learning. Education is primarily to develop individual knowledge and intellect, whereas the purpose of training is to develop a specific skill. Using this criterion, the training for the Board of Directors is actually more likely to be awareness or education.

The type of training that is appropriate will depend on the availability of the trainees. It may not always be easy to release the trainee from their job to attend a classroom or training centre. For example, if the trainee is a process operator on a large refinery, it may be better to use online training that they can access at a time to suit themselves when the refinery is running steadily and requires less of their attention. Likewise, people who are lone workers, such as security staff, may find it difficult to be released for "off the job" training. There is no standard formula to define where training should take place. It all depends on the constraints that exist in the operation of your facility. However, do not select a training approach that will exclude some of the target audience, as this will undermine your credibility and will create extra work later on.

Generally, I find that for most people at the shop floor level, the closer that the training takes place to their own workplace, the more relevant it will seem and the more interested and engaged that they will be. For them, they will normally prefer training in the workshop and this is fine, provided that no specialist training or simulation equipment is required and that the number of trainees are very small. On the other hand, I find that with managers and professional staff, there is a need to get them away from the workplace and the hurly-burly of running their operations. There is a tendency for managers to be continually answering phone messages or checking their emails, and this can seriously disrupt their involvement in the training event.

One of the international Health and Safety Management workshops that I have run for the last 15 years is intentionally run in a hotel/conference centre for exactly that reason. We found it was the only way to get managers to concentrate on the subject in hand. We also found that delegates at the workshop were frequently asking to leave the training sessions halfway through the last day in order to catch a flight home. This is extremely disruptive to the event and has a tendency to spread like wildfire. The general feeling is that, if one person doesn't need to attend the last half-day, then why can't they all leave early? The solution to this problem was to set up a very high-profile "Course Dinner" at the end of the last day of the workshop. This dinner was hosted by a very senior executive of the company, and gave an opportunity for him to convey his personal commitment to health and safety. The dinner was also the time at which various awards were presented, and in general it was an excellent social and networking session. The key point, however, is that attendance at the workshop was made conditional on the delegate attending the course dinner at the end of the event. Admittedly this approach did add a little to the overall cost of the workshop, but the result was that the quality of the training itself was greatly enhanced, purely by addressing some of the important "hygiene" factors.

Issues relating to the availability of trainees are not limited to shift or lone workers. I frequently get asked to educate very senior executives about health, safety and environmental matters. I often find that it doesn't matter how broad the training need may be, the time that Board Members are able to commit may be very limited. In these circumstances it is necessary to be creative in how the training is applied. Instead of inviting the Board of Directors to spend a week off the job at a health and safety training course, a more acceptable approach may be to ask them to invite you to provide an hour's input before the normal business of their monthly Board Meeting. This drip-feed approach is invariably more acceptable to senior executives and has the benefit that once established as a part of the normal Board Meeting process, it can become an ongoing feature, resulting in never-ending improvement at the Board level.

In summary, when preparing for the training there are six steps that need to be considered:

Step 1. What are you trying to achieve? What is the objective of the training and what is the topic?
Step 2. Is the training required at the specialist, competence or awareness level?
Step 3. Is there an immediate need (i.e. can the training be applied immediately)?
Step 4. What is the availability of the trainees? Can they temporarily be released from their responsibilities, or does the training need to be tailored to fit with their working pattern (e.g. shift working)?
Step 5. How much time is likely to be available for training?
Step 6. Is there a qualified trainer available?

"Audience" Checklist:

1. Understand why they are there. Are they:
   - Voluntolds (Pressed men!)
   - Vacationers (Glad to be off the job!)
   - Valued learners

2. Suit the training style to the needs and capabilities of the individuals.
3. Understand the personality traits that you might meet during the training and how to deal with them.
4. Remember to:
   - Tell them what you are going to tell them
   - Tell them
   - Tell them what you told them
5. Ask the trainees to play their part:
   - Enter into the discussions ENTHUSIASTICALLY
   - GIVE FREELY of your experience
   - CONFINE your discussion to the topic
   - Say what you THINK
   - Only ONE PERSON should talk at a time
   - LISTEN CAREFULLY to what is being said
   - BE PATIENT with other trainees
   - APPRECIATE the other trainees' points of view
   - BE PROMPT
6. Pay attention to the ergonomics of the training session. Use the most appropriate layout that will aid interaction between the tutor and the trainees.

Chapter 5

Training needs – What do they need to know?

The ancient Chinese philosopher Confucius' sayings are still relevant today. He reminds us that we should "*Give instruction only to those people who seek knowledge after they have discovered their ignorance*". In other words, focus your training in those areas where it is needed.

My own experience while studying Engineering at university many years ago was that we had some brilliant lecturers and tutors. The problem was that they were so knowledgeable about their subjects that they found it difficult to understand why mere undergraduates could not grasp some of the complex concepts that they found so easy. Knowledge and intellect alone do not necessarily make one a good teacher. To be a good teacher in the work environment, one needs to have empathy with the people you are trying to train. In particular you need to put yourself in the trainee's position and understand what they need to know and have the ability to demystify the subject to a level that can be understood by the trainees who are present on that day. It is equally important to recognise what they do not need to know at that particular time. Teaching people information that they will not be using in the immediate future is likely to de-motivate them and once demotivated, their attention wanders and they fail to absorb even that information that is immediately relevant to their needs. I have attended very many safety training sessions over the years, and a common mistake is that the tutor attempts to bring the trainee up to the tutor's knowledge level. In most situations this is unnecessary and will flood the trainee with a lot of information much of which they will not need or be able to use.

5.1 IDENTIFYING TRAINING NEEDS

Before starting to design the training, it is important that there is a clear need. This need for training can arise from a number of different sources. In some circumstances the training content may be mandated by law and the training curriculum, format and validation process may be prescribed. In these cases, it might be necessary to go to an externally accredited training

DOI: 10.1201/9781003342779-6

provider. If the training need is recognised internally within the organisation, it might have arisen as a result of a recommendation from an investigation into some sort of accident or incident. Alternatively, it could arise because of a change in normal working practices. This might be as a result of the introduction of some new equipment or changes in standard procedures or works instructions. Training needs can be identified during individual's annual personal appraisals or from personal requests. Finally, it may be that the organisation has done detailed training needs analyses for each job which identify what competencies are required for each job and where the gaps lie between the required competencies and those exhibited by the job holder. So, to summarise, the training need can arise from:

- Mandatory training requirements laid down by law
- New recruitment
- Changes in personnel responsible for the activity
- Recommendations arising from an incident
- Changes to procedures and practices
- Introduction of new or changed equipment
- Recommendations from annual appraisals
- Training needs analyses
- Changes in external regulation or standards
- Bad habits and shortcuts

Training is required when there is both a need and where the individual does not currently have the competencies to fulfil that need. Having identified the requirement for each job role or individual, this can then be compared to the current skills and knowledge that each person in that role has. An individual's current skills can be identified in a number of ways and could arise from:

- Documented proof of previous training (e.g. Certificates, examination results, etc.)
- Existing training records
- Discussion with the individual
- Proven track record
- Appraisal discussions
- Knowledge checks

When identifying need, an important issue that a tutor should be aware of, is that sometimes the "need" is not actually a training problem. It may well be that the student already knows the right way to do something but failure to respond in a safe and appropriate manner is caused by other factors. For example, the student has not been provided with the right tools, or perhaps there is a disciplinary issue. The tutor may need to alert management to the fact that they will have to address the factors that are preventing people from following the process or procedure. This is not an easy thing to do, but it is an ethical responsibility to at least give management information.

5.2 KNOWLEDGE CHECKS

Once the requirements of a particular role or task are identified, the knowledge check is produced by creating a series of questions that are clearly related to the task and that give the participants the opportunity to demonstrate their competencies. The questions should not be trivial but be reasonably challenging. The knowledge check should not try and trick the participants. Remember that the objective is only to get an understanding of the current knowledge and skill level of the participant and is not about finding fault or identifying weaknesses.

Once the current knowledge level of an employee is known, that can be compared with the skill requirement for the job. If there is a gap between the two, then a training need has been identified and arrangements should be put in place to satisfy this need. The use of Training Needs Analysis is a very important part in customising your training and making it relevant to the individual and economically beneficial to the organisation.

If the knowledge gained from training cannot be applied almost immediately, then it is likely that the trainees will not understand its relevance and they will have forgotten what they have learned before the opportunity arises to practice their new skills. It is also very difficult to establish competence if the training is not followed quite quickly with some monitored application of the training leading to validation.

5.3 TRAINING LEVELS

Once a clear training need has been established, then it is important to establish at what level the training is required before it can be planned and implemented. In the work environment, there are usually three broad knowledge levels to consider. These are:

Level	
SPECIALIST	A specialist has an in-depth knowledge of the subject at an expert level
COMPETENCE	Competence is the ability to carry out a task to an effective standard. To achieve competence requires the right level of knowledge, understanding and skill, and a professional attitude.
AWARENESS	Awareness is general familiarity, of a subject

It is very important to establish this criterion at a very early stage, as the time commitment, number of trainees, scope and tutor skills will differ markedly if it is "Awareness" level as opposed to "Specialist" level training. If we consider the situation in the average factory, not everyone needs to have detailed knowledge of every piece of legislation that is relevant to that facility. Typically, that detailed knowledge may be restricted to two or three people on the site. However, everyone will need to be aware of the bits of legislation that are relevant to their particular role. In these circumstances, the person with the specialist knowledge should interpret that for everyone else and make sure that the relevant requirements are incorporated into the local procedures and instructions. Very often detailed regulatory knowledge is held within the Health and Safety Department, and these individuals will need to be trained and educated to a very high level in the details of regulatory requirements. This requires detailed "specialist" level training.

Typically, members of management are responsible for ensuring overall compliance with health and safety regulations. These individuals do not need to be health and safety specialists, but they need to know enough in order to be competent managers. A competent manager needs to know what his team should be doing and to know enough to recognise when they are not complying with the rules. Furthermore, he or she should be sufficiently knowledgeable to recognise the limitations in their own knowledge and when to seek further specialist help. The most dangerous type of manager or worker is the one who doesn't know when to stop. The problem is that these people "Don't know what they don't know!" In these circumstances, these people need to be taught sufficient to ensure that they remain competent.

The third level of training is that of "Awareness". This level will tend to apply quite widely across the workforce, but it is done in much less detail than with the "Competence" and "Specialist" levels. This is the level of health and safety training that most people will experience for much of the time at work. Awareness training is the most challenging in terms of trainee engagement. Specialists are largely self-motivated and frequently are responsible for identifying their own shortcomings and the subsequent need for additional training. Likewise, managers recognise that in order to pass on their knowledge to their teams, they need to be competent and understand the subject themselves. In the case of the large raft of "Awareness" training, the trainees will largely see this as being imposed by management and so they are often initially attending this with a rather negative attitude. This provides a particular challenge to the tutor, who needs to get the trainees engaged and interested in the subject. This is especially a problem with health and safety training. It is not like task training, where the individual needs to pay attention in order to learn how to do their job. Health and safety training is often seen as boring and statements of the obvious or not directly relevant to one individual's performance at work. The challenge that most health and safety tutors will face is likely to be in the area of "Awareness" training.

5.4 RECORDS

Maintaining up-to-date training records is a very important part of the process of any individual's competency assurance and is a management responsibility. It may seem like an unnecessarily bureaucratic task, but if something goes wrong, the first thing that the regulatory inspector will ask is to see the training records. If the records are incomplete, the inspector may initially assume that the individual was untrained, and it will be up to the management team to try and prove otherwise. Records are particularly important when it comes to proving that the employee had been told the right way to do something and on the occasion under investigation, he or she had not followed their training.

Training records not only need to record the fact that training was completed on a particular date, but they should also be capable of cross-referencing to the content of the training, so that the details can be established of exactly what the individual was trained to do and how it should be done safely.

Training records are an invaluable part of any training needs analysis.

Training Needs Checklist

I. Is there a real need – i.e "why are we doing the training"? Is it because of:
- Mandatory training requirements laid down by law
- New recruitment
- Changes in personnel responsible for the activity
- Recommendations arising from an incident
- Changes to procedures and practices
- Introduction of new or changed equipment
- Recommendations from annual appraisals
- Training needs analyses
- Changes in external regulation or standards
- Bad habits and shortcuts

II. What level of training is required?
- Awareness
- Competence
- Specialist

III. Establish the current level of knowledge by:
- Documented proof of previous training (e.g. certificates, examination results, etc.)
- Existing training records
- Discussion with the individual
- Proven track record
- Appraisal discussions

IV. Ensure good records are kept

Chapter 6

Designing the training – What do they need to know?

Too often we make our objective "delivering training" instead of "delivering the right training to the right people at the right time". Before we can invite anyone for training, we need to have a clear scope for the training. For example:

1. Who is the training for?
2. How many people need to be trained?
3. Has something happened to trigger the need for this training (i.e. Is it urgent)?
4. At what level is the training required? (Awareness/Competence/Specialist)
5. Do we have the skills to deliver the training in-house or do we need to buy it in?
6. Will the training be in a group or one-to-one?
7. Will it be on-the-job, classroom or internet/intranet based?
8. How many people need to be trained?
9. Do we have suitable training facilities?
10. Do we have a suitably competent trainer (technically and training skills)?
11. How will we evaluate the success of the training?

6.1 SESSION SPECIFICATIONS

If there is no readily available training programme, then one will need to be produced. If I am starting from scratch, I like to develop a Session Specification. The purpose of such a specification is to clearly identify the purpose of the training and a framework of how it could be delivered. For example, who will do the training, how much time is available and needed, and what sort of practical work and exercises might be needed.

An example of a Session Specification might be as follows:

DOI: 10.1201/9781003342779-7

Session Topic:

Health & Safety Roles & Responsibilities

Trainees (Roles & Numbers): First Line Managers - 15

Duration 90mins Day 1: 0930 -11 00	Presenter: Simon Pain

Scope (Content):

- Who is accountable?
- Management liabilities
- Management behaviour
- Knowing your limitations
- Competence / Control / Co-operation / Communications.
- Link to HS(G) 65
- What should "enlightened" management be doing?

Style: (For details see following page)

- Team building Exercise. (Jigsaw). - 15mins
- Keynote presentation - 30mins
- Individual Exercise. (Questionnaire). - 10mins
- Group Task. (3 separate tasks). - 20mins
- Group feedback & discussion - 15mins

Group Task / Activities:

What makes a good leader?

a) Motivating. (Group 1)
b) Setting & Maintaining Standards. (Group 2)
c) Developing others. (Group 3)

Relevant Standards:

Corporate Standards EHS101 / 106 / 107

Assessment method

End of course personal plan

Facilities:

Training Room suitable for 15 in U-shape

3 syndicate rooms suitable for 6 persons with boardroom style layout

Figure 6.1 Example of training session specification.

The purpose of such a draft specification is to ensure both the "What" and "How" of the training is properly considered and agreed before anyone spends a lot of time developing the detail of the session. It is important to ensure that the training design covers all the training requirements and that it will be done in a way that has the greatest chance of being successful. Once the training designer has completed a draft of the specification, this should be shared with the line manager (or whoever requested the training) and also with one or more of the potential trainees, to make sure that the training will meet the needs of the trainees and their management (Fig 6.1).

Once the specification is agreed, then approximate timings should be allocated. Timings need to take account of the fact that many trainees may not be used to sitting for long periods of time and remember that your brain can only absorb what your backside can endure! Although training should take whatever time is required to be successful, it is common for managers to think in training blocks of an hour, half a day, a day or multiple days. The higher up the management pyramid you go, then the less time will tend to be allocated for training. The most challenging group to train in health and safety are the most senior members of the organisation. In these situations, the best solution is to seek a regular short slot in board meetings or the like and use those to educate the senior executives (see Chapter 5).

An example of how to allocate timings is shown below. It should be recognised at this time, that as some of the content may yet to be developed, these timing should just be initial targets. As a guide I find that on average I spend about 3 minutes/slide of a presentation, but you may be different and the only way to be sure is to practise the actual finished presentation.

My experience is that most people are much too ambitious in how much material they think that they can get through in a given time. Practical exercises and syndicate working are very useful in raising interest levels and involvement, but do remember that they do take a lot of time. In planning a syndicate break-out exercise, remember to allow for the time required to explain the exercise, the time lost for getting to and from the break-out rooms, the time for the syndicate work or exercise itself, the time for each of the groups to feedback and finally the time for the tutor to summarise the learning (Fig 6.2).

If the training session is to be one of a series of similar or identical events, then as mentioned previously, I recommend doing a trial run on a one-to-one basis with a colleague who understands the subject and who can provide constructive feedback. Remember to ask your colleague to ask salient questions during the trial run because this is what you can expect when doing the training for real. It is sometimes advisable to run a "pilot" training session once you think that you have developed the training material. In this case, be selective about who is invited to the pilot session. There are always some trainees who either don't want to be there (the Voluntolds!), or who think that they know it all already, or who just want to be disruptive; these are not the types to invite to the pilot. When inviting people to the pilot session, make it clear that it is a pilot and that at the end

Proposal for Safe Systems of Work using the Daily Work Control Permit & Point of Work Risk Assessment

Time	Topic	Key elements	Exercises / Features	Exercise location
10 mins	Introduction	Intro by a Senior Manager		
Section 1 **The principles of Risk Management**		**Built around the 6 steps**		
30 mins	Step 1 – Hazard Identification	– Introduce the 6 -tep process – Understanding difference between Hazard & Risk – Variability of consequences	Hazard spotting exercise	Plenary (Work in pairs)
10 mins	Step 2 – Who / What can be affected	– Unintended consequences – Contractors / other workers / visitors – Special cases (Young people, expectant mothers) – The 3 "P's"	3-legged stool	
75 mins	Step 3 - Risk Assessment (Coffee break during this session)	– Qualitative & Quantitative risk assessment – Purpose of a risk assessment – Legal duties HASAWA Regs 7 & 9 – Risk Perception – Positive & negative motivators – Suitable & Sufficient risk assessment – The STSC Risk Matrix – ALARP – Criteria – Other types of R.A. (COSHH, VDT, Manual Handling)	Risk perception flashcards exercise Consequences Game Risk Matrix exercise	Plenary (Groups of 3) Plenary Plenary
20 mins	Step 4 – Workplace Controls & precautions	– Concept of Residual risk – Difference between Acute & Chronic risk exposure – Hierarchy of Controls – Risk effect of new controls		
10 mins	Step 5 – Record & Communicate findings	– Importance of recording the R.A. – Who to communicate to? – Do they understand?		
5 mins	Step 6 - Review	– Broader meaning of the concept of review		
25 mins	The Risk Game	Fun activity that Reinforces the 6-step process (Note: this activity takes about 25 minutes and could be removed if the course needs to be shortened. However, it is usually well received)	Risk game	Break out rooms Max of 6 persons / game
		Break		

Figure 6.2 Example of programme for a training event.

you will be interested in their constructive feedback and suggestions for improvement. However, if you do run a pilot, it is important to be seen to listen and respond to the feedback being given! The best example that I had was when I was invited by a client to develop and run a series of health and safety auditing appreciation courses. My client contact had asked for the course to be run in a day. Following the pilot course, the trainees unanimously recommended that the course should be made longer to allow them to fully appreciate and practise the learning. That is quite unusual – trainees nearly always want to shorten the training!

Trainees will forgive their tutor almost anything, but from their point of view, what is completely unacceptable is for the tutor to over-run the allocated time. This is not only a problem for the trainees, who might need to get away to catch a bus or collect their children from school, but if the training session is a part of a larger course, it may affect the following tutor or if the course is run in a hotel or commercial conference centre, it might adversely affect arrangements for refreshments or lunch. Once trainees get the impression that you are going to over-run your time, their attention will be affected by an irritation factor and so their understanding and appreciation of the training information will diminish significantly. One tutor that I observed during a training session had used up ¾ of the available time but had got through less than ¼ of his presentation. By a herculean effort he managed to display all his remaining slides in the last part of the session. I noticed that all the interactions that he had in the early part of the session disappeared and the trainees just wanted it all to end. Although he finished on time, I suspect that the trainees registered or remembered very little of the latter part of his presentation.

It is important to have clear timings identified during your training, so that you can recognise a potential timing problem early enough to react and do something about it. The early warning signs are if there are substantially more questions than you had expected. This is a balance. Good training stimulates involvement and involvement is often materialised by questions being asked. The challenge comes when there are so many questions that it starts to affect the timing of the session. If the training is on a one-to-one basis or there is no difficulty in deferring some of the training to another day, then take as many questions as are necessary, however, if the session has time deadlines, then it may be necessary to curtail the questions or curtail some of the "sacrificial content" (see section 6.2).

Ways in which you can curtail questions without de-motivating the questioner might be to say:

> *"That's a great question, but as time is getting on, would you mind if I discuss that with you later?"*

Or if you want to stop further questions on this particular subject you might say:

> *"If it's OK with you all, and in the interests of getting you all away on time tonight, can I suggest that we move on?"*

I often get involved in designing or approving presentations for inexperienced presenters. Keeping to time is always a challenge. It is an interesting human behaviour that although I always have a large clock in view of the presenters, they seem to not register what the clock is telling them. To overcome this time-keeping problem, I have developed a count-down system. With the agreement of the presenter, I sit at the back of the training room in direct sight of the presenter but not in the direct sight of the participants. I have made an A4/Foolscap-sized flip-top pad with various messages in large type. Halfway through the allotted training time, I would display a "½ way" sign. I have found even then, if the presenter is really engaged in what he or she is doing they may not notice, and so I wait until I get a nod from the presenter to show that they have registered the message. It is important not to overdo this and so typically my next warning would be "½ Hour left", then "15 Minutes" and then finally the last 7 minutes would count down 1 minute at a time. I also have a "STOP" sign in the event that the timing goes completely wrong and I can see that it is having adverse effects. I have only ever had to use that once!

6.2 SACRIFICIAL CONTENT

When it comes to short-cutting your training presentation, it looks really unprofessional to bring up the slide-sorter page for everyone to see and witness you select and miss out a series of slides. Trainees might feel cheated and be left wondering what it was that you missed out. If you inadvertently find that you need to skip some slides using the slide sorter page on your computer, then do it during a natural break with the screen projection switched off, or while the trainees' attention is distracted by working on an exercise. In preparing the presentation, think in advance about whether there is any part of the presentation that is less important and put in a discrete hyperlink to a future slide. Make sure that the link is initiated by some unobtrusive and unidentified button or symbol somewhere near the bottom of the slide, so that you don't give the game away. However, remember that if you have provided detailed handouts showing the slide material that you intend to present, the trainees may realise what you are doing, if you miss out some slides and that may cause more confusion than it helps. It is usually better to use some exercise, video clip or break material as your sacrificial content, if time could become a problem. However, remember that usually you will not know whether you are running out of time until you are into the later stages of your training – so the sacrificial content will need to be towards the end of the session.

If you are running a training course involving multiple sessions in the same day using different speakers, then it is essential to provide both speakers and trainees with a programme showing the detailed timings. The problem is that once you have publicly revealed your timings, you are committed and need to meet them. However, if you are the sole tutor and the session is intended to go on for a full day or more, then to avoid some of the time management problems, I recommend identifying the key stages in the training, but do not commit publicly to the precise timings in the day when each stage will be completed. You will of course need to have your own private timing targets, but this approach gives you more flexibility to manage the time over and under runs during the day. In this situation it is still important to tell the trainees precise timings for lunch/coffee breaks and the end of the day and to make sure that you achieve those.

If you have more information that you can cover in the allotted time, then I would normally recommend making additional time available. If this is not possible, then additional "reference" material can be provided and appended to the back of your handouts. Just be aware that human nature will tend to dictate that the chances of this material ever being read is low and so its training value is not very effective!

Finishing training a few minutes early is usually appreciated by attendees, but finishing an hour of two early looks unprofessional and might suggest that you don't know what you are talking about. The key messages around training time management are:

1. Don't try to cover too much
2. Structure the training so that it develops in a logical manner. Speaking "off the cuff" is very likely to lead to time over-runs
3. Practice your presentation beforehand
4. Do not allow yourself to wander off the topic. Stay "on message"
5. Decide in advance how you might manage time over-runs (have "Sacrificial Content")

Training often takes a lot of time commitment for both the tutor and the trainee. As you are going to be spending quite a lot of time together, it is important to address some of the "hygiene" factors about where the training is to be done. Sometimes there is little choice if you need to be close to some piece of fixed equipment that is associated with the training, but quite often health and safety training is done within meeting rooms and the selection of the right room can affect the success or otherwise of the training.

Designing the Training – Checklist

Decide why the training needs to be done:

- Who is the training for?
- How many people need to be trained?
- Has something happened to trigger the need for this training i.e. is it urgent?
- At what level is the training required? (Awareness/Competence/Specialist)
- Do we have the skills to deliver the training in-house or do we need to buy it in?
- Will the training be in a group or one-to-one?
- Will it be on-the-job, classroom or internet/intranet based?
- How many people need to be trained?
- Do we have suitable training facilities?
- Do we have a suitably competent trainer (technically and training skills)?
- How will we evaluate the success of the training?

Produce and agree a programme/framework for the training (Session Specifications)

Avoid the pitfalls:

- Don't try to cover too much
- Structure the training so that it develops in a logical manner. Speaking "off the cuff" is very likely to lead to time over-runs
- Practice your presentation beforehand – if necessary do a trial run
- Do not allow yourself to wander off the topic. Stay "on message"
- Decide in advance how you might manage time over-runs (have "Sacrificial Content")

REMEMBER THE GOLDEN RULE:

DO NOT OVERRUN YOUR TIME ALLOCATION

Chapter 7

Training styles and techniques – Making the training interesting

7.1 WHAT TYPE OF TRAINING IS APPROPRIATE?

Once training needs have been identified, there are a multiplicity of possible training techniques available. Not all techniques are suitable for all training needs. It is important to decide whether the need is for general education, specific skill training or for communication. The techniques used will differ depending on that initial objective. Some techniques are more appropriate than others for different purposes or different groups.

The following table (Fig 7.1) is from Professor John W. Newstrom's publication "Evaluating Effectiveness of Training Methods". It should only be used as a guide but might help clarify what sort of training technique is appropriate in what circumstances. What is interesting from this work is that the best approach for knowledge acquisition is not necessarily the best approach for knowledge retention. Games (see Resources Section H) and role play (see Resources Section I), which both

Ranking of methods depending on Teaching Goals (1=high; 8=low)

METHOD	Knowledge Acquisition	Attitude Change	Problem Solving	Inter personal skills	Participant Acceptance	Knowledge retention
Case Study	4	5	1	5	1	4
Workshop	1	3	4	4	5	2
Lecture	8	7	7	8	7	3
Games	5	4	2	3	2	7
Films	6	6	8	6	4	5
Role Play	2	2	3	1	3	6

Source: John W..Newstrom, Professor Emeritus of Management in the Management Studies Dept. of the Labovitz School of Business and Economics at the University of Minnesota Duluth

Figure 7.1 Evaluating effectiveness of training methods.

DOI: 10.1201/9781003342779-8

usually have a high level of involvement and fun tend to be very memorable with trainees.

In particular, careful thought needs to be given to the situation where trainees cannot be easily released from their workplace because of shift working constraints, lone working or travel requirements. It is also necessary to decide whether the organisation has the skills or facilities itself to carry out the training or whether there is the need for some external accreditation (e.g. for High Voltage Electricity switching). The table shown in Fig 7.2 below can be used as an initial guide to assist in what type of training is likely to be suitable for a specific training need.

The selection of training techniques may affect the decision as to how and where the training will be carried out. Every technique has a balance of advantages and disadvantages. Circumstances can dictate that a different approach may be needed at different times. For example, periodic refresher training might be done via an interactive computer training package, whereas the initial training on the same subject might require more detailed face-to-face training.

The (Fig 7.2) identifies some of the benefits and disadvantages of different training techniques.

In our earlier discussion in Chapter 2 about memory, we learned that the brain has two sides which have rather different functions. In the Top 20 list we also see that point 9 reminds us that people remember more when they both hear and see information. One of the reasons for this is that the function of speech comes from the left-hand side of the brain and images, colour, expressions and comprehension are processed on the right-hand side of the brain. Using both sides of the brain at once enhances its capability. Consequently, this means that using both senses of hearing and sight concurrently results in better understanding and better memory retention. People only tend to remember about 10% of what they hear and 20% of what they read. However, if trainees both hear and see information at the same time, they can remember up to 80% of it. If we are both listening and seeing information at the same time, we are using both sides of the brain and the result is greater than the sum of the two parts operating independently. This is why training should involve talking about and seeing information – and if possible being able to put that learning into practice in a safe environment (Fig 7.2).

The visual side of training is the easiest in which to create impact and make your message memorable. There are many ways of creating visual impact. The most commonly used and well-known visual presentational aid is the projected presentation. The benchmark system is Microsoft PowerPoint, but there are other commercially available systems that have some special functionality that might make them more appropriate if you are preparing computer-based

Training Format	Applications	Advantages	Disadvantages
Lecture	Mainly one-way communication • Factual knowledge transfer • Policy communications • New legislation • Accidents & incidents	-Large audience -Common message -Good for Knowledge transfer	-Little interaction -Tutor cannot easily judge reaction Not good for skill transfer
Training Course	Bespoke Skill training Complex message	-Workgroups -Good interaction -Q&A possible Good for skill transfer	-Time consuming -Off the job
Tool box talk (i.e. Brief work team get-together on the job)	Communication of standard messages that are relevant to the group. (e.g., Accident prevention) Suitable for a natural work group	-Message is relevant to that workplace -Short -Q&A possible -Group leader seen to own the message -Good for bespoke communication	-Tutor is usually the group's team leader or manager who may not be skilled in training
Videos / DVDs	General subject matter Suitable for Shift or lone workers	- Easy to arrange -Good range of commercially produced videos -Produced by experts -Any size audience -Good for conveying general principles	-May not exactly fit the local requirements -Not easy to deal with questions -Easy not to pay attention -Rarely any validation
Interactive / online	General or workplace specific subject matter Suitable for Shift or lone workers or other individual training.	-Either commercially available or home made -Easy to arrange -Usually professionally produced. -Good range of commercially produced interactive training available -Some can do a Level 2 validation check -Can be done in stages	-May not exactly fit the local requirements -Not easy to deal with questions -Easy to cheat!
On the Job	Transfer of job specific skills or knowledge required	-Easy to arrange -Validation possible -Suitable where there are a diverse range of different tasks	-Short cuts & bad practice can get passed on -Safety standards can be diluted over time -May not always use a good trainer
Reading	Job Instructions & procedures Legislation Suitable for individuals	-Cheap -Accurate	-Not a good way for individuals to retain knowledge -Some people don't like reading. -Validation is essential

Figure 7.2 Advantages and disadvantages of different training techniques.

training. However, there are also other forms of visual presentation that can also be used. The sort of visuals that can have impact on the trainees are:

- PowerPoint presentations
- Videos
- Models
- Posters or chart boards
- Sticky note displays
- Real-life experience
- Simulations
- Role plays
- Photographs
- Jigsaws
- Drawings and charts
- Gizmos
- Exhibits and displays

We will look in some detail at these in the next two chapters.

7.2 VIDEOS AS A TRAINING AID

The use of videos for training is quite widespread. I find that participants either love them or hate them. There are several different ways of using pre-recorded film. The most common and easiest for the tutor is to use commercially prepared "professional" videos. The other alternatives are to produce a bespoke video yourself, to use short video clips or to use video capture, where you video record the trainees doing some task and then play it back as a form of critique.

7.2.1 Commercial Videos

Advantages

- Usually well presented with a clear and logical script.
- Usually will have quite high impact.
- Can be memorable.
- Good graphics.
- Easy for inexperienced trainers to use.

Disadvantages

- The material does not always match your situation very well.
- They are often long.
- It is not easy to ask questions.
- They can be expensive to purchase (consider hiring?)

My own experience is not to use many commercial videos. However, if they are a good match for your needs, they can reduce your workload. They can also be a good facility for team leaders, supervisors or inexperienced trainers in situations such as toolbox talks or during night shifts, when there is no other training support available. If commercial videos are used, then make sure that time is allowed at the end to discuss the content of the video and to relate it to the situation in the trainee's particular workplace.

A particularly good use of commercially produced videos is where there may have been a major incident elsewhere in your industry or the world, where you believe that there are learning messages for your team. Showing a professionally prepared video of what actually happened and then following that with a discussion about "whether a similar situation could apply here" can have high impact. Some excellent videos and video simulations of real events are available free of charge from the US Chemicals Safety Board website (https://www.csb.gov/videos/)

7.2.2 Bespoke Videos

Most people are not professional film producers. The danger with homemade videos is that they can become a bit of an opportunity to poke fun at the amateur actors and to find fault with the video production, rather than focusing on the content. On the other hand, a bespoke video that is well produced can be very powerful. I was helping one client with some training for an international conference, in how to communicate health and safety effectively. We wanted to make the session very interactive despite the fact that it was at an internal company conference with over 80 people there. Having trained the group in a number of safety communication techniques, we wanted to let everyone take part in a simulated, but realistic safety communication exercise. We had set up in advance a number of role play scenarios using professional actors. Each scenario showed a workplace situation where there were various unsafe conditions and unsafe acts exhibited. One location was an office, one was a laboratory, another was a production area and the final one was in a warehouse/store. Each of these scenarios was video filmed in advance. A different actor took the lead in each filmed scenario. During the training, the group was broken into five sub-groups, with each sub-group watching a different video. Having watched the scenario, the actual actor in that video appeared and was interviewed by one of the trainees. The results had very high impact when some of the trainees got a little too demanding, resulting in one of the actors breaking down in tears (all pre-planned of course!) After a feedback session, the trainees moved round to each of the video scenarios, in turn, allowing a different trainee to experience first-hand the consequences of good or not-so-good behavioural safety communication skills. This approach was deemed to be very realistic and had such high impact that people still remember it five years later. There are a few barriers to this approach. It is expensive and requires a lot of preparatory time. Using actors in training situations is not unknown, but it is important to use actors who have some understanding of your type of workplace. The actors will require a very detailed brief for each scenario, so that they can answer questions in a realistic manner. Don't forget that actors may not be aware of industrial-type hazards. If you are simulating unsafe acts and conditions for video demonstration purposes, you must always do a risk assessment before filming. (A full explanation of this technique is detailed in the Resources Section D at the back of this book.)

It is also quite effective to just use short video extracts to make or reinforce a particular point. Remember if you are taking a clip out of a commercially produced video you may be in breach of copyright regulations. The best way of doing this is to use short clips that you have taken yourself to demonstrate a particular hazard or breach that you have seen at your facility. There are also some websites that allow safety trainers to use their material copyright free. As mentioned previously, one of the best of these is the website of the Chemical Safety Board of the USA. That website has a series of videos explaining the causes of various major incidents that have occurred in the United States. The videos are quite long, but many of them have excellent animation sequences explaining pictorially what happened, and these animated extracts of the video form very powerful messages for many different health and safety training sessions. The CSB encourages the use of these videos or extracts from them.

I have also used some clips from various TV "blooper" shows. These usually give a light-hearted break in proceedings. If you are going to do this, remember that you are not just trying to entertain – there must be a purpose to the clip. One clip I particularly liked came from the Netherlands and featured an attractive young lady (an actor) passing various people (unassuming members of the public) on a moving escalator. As she passed each man, she gave him a kiss. The first was to an elderly gentleman and judging by the expression on his face, she made his day! The final kiss was to a young man who was with his female partner – the male seemed to be surprised, but his partner obviously was not amused. This clip always raises a laugh, but the point of it from a training point of view is the message that you can do the same thing to different people and get totally different reactions. It also reminds us that when doing health and safety training we may need to adapt our approach for different people or audiences.

Videos can also be used very effectively as observation exercises. People tend to think that they are good at "seeing" things because after all, they have been doing it all their life! My experience is that people often do not see hazards in their own workplace because they get used to accepting that things are always like that. A good application of video training is to take a video of workplace conditions and people working in your own work area. This can then be played back to get people to identify hazards and what suitable control measures could be introduced to minimise risk. Video clips can also be used to demonstrate how our observations are influenced by our expectations. I use various humorous well-known clips from comedy programmes – my favourites for this purpose are sketches from "Monty Python's Flying Circus" such as the "Parrot Sketch" or the "Lumberjack Sketch". Showing these video clips with minimum introduction leads the participants into believing that this is just a light-hearted break in proceedings. At the end of the clip, I then ask the participants to answer a few questions about what they saw in the sketch such as "What was hanging from the ceiling?" or "What was the main character (John Cleese) wearing?" Because the participants were engrossed in the sketch itself, they paid little attention to the background and so they rarely get any of the questions right. I then go on to show a different sketch, but this time I tell them in advance what I want them to look out for. Although they still enjoy watching the sketch, they invariably get all the questions right because they had warning about what I wanted them to answer. The point of this exercise is that if you want to train people to use their eyes to spot hazards, then it is no good telling them to go and have a walk around their area and see if they can spot that anything is wrong – you need to provide them with a focus to look for. So be specific, it may be that you go and tell them to see if there are any machinery guards missing, or whether all scaffold boards are safely attached to the scaffolds.

A particularly good use of short video statements is to ask the Managing Director or other senior person, to say a few words on video that can be used as an introduction to the training. This emphasises the health and safety commitment of senior management and shows that this particular training is important. If you do use a video introduction in this way, then it is important that the comments are relevant to the training topic and that it is specifically used for one particular event or repetitions of that training event, and is not just used as an introduction to any and every training event.

Finally, another very powerful way that you can use videos in a training environment is to get people to talk on camera about some of their own personal learning events. This is usually much easier for the victim than coming into his own workplace and having to talk live to his or her colleagues. People who have had significant injuries later often recognise that their own actions in some way contributed to their injury. Some of those people may be willing to be videoed, talking about what went wrong and what they think other people could learn from their actions or omissions. If you are thinking about doing this, then remember that it is not a good idea to do it too soon after the incident, as the victim will need time to recover as much as possible and to have time to reflect on his or her contribution to what happened. They will also need to be reasonably articulate. For shop floor training, you will find that open and honest feedback from an injured colleague is a much more powerful learning event for trainees, than being told about it from someone like a manager who was not directly involved. A particularly powerful series of commercially available videos from Latitude Safety Ltd have been made by an individual who worked at a soft-drinks manufacturing company. This individual was blinded by the accident and has now dedicated his life to talking to companies about his experiences and how repeat events can be avoided. One of the videos that I would recommend is "Hindsight – The Official Ken Woodward Story" (Fig 7.3).

Figure 7.3 Ken Woodward.

Source: Courtesy of Latitude Safety Ltd.

The one caveat in the approach of producing in-house videos recounting a victim's own learning messages is that you must be prepared to show the video "as recorded". If you edit it and that changes the impression of what was being said or to protect someone else, then the value of the video will be undermined and you may be accused of a cover-up. If you feel a need to influence the injured person's video feedback, then you can always set it up as a TV-type interview, where the interviewer poses specific questions to the injured person. In these situations, I would normally have the interviewer "off camera".

7.3 MODELS

We have discussed the importance of visual images in reinforcing understanding and memory. Sometimes it can be helpful to not just use a photograph or picture, but also to use a model. The more different and original types of visual image that you use, the more likely trainees are to remember it. I tend to use models when I want people to remember a concept.

Safe working is not only about having the right plant or equipment. You can have all the best tools and equipment in the world, but if people don't know how to use them, then unsafe situations can arise. To get trainees to understand the importance of this concept I use the 3 "P's" acronym. This means that in order for things to be safe, we need to ensure that we have considered and applied the safety requirements arising from

- PEOPLE
- PLANT (and equipment)
- PROCEDURES

I liken this concept to a three-legged stool. A client of mine was very keen on personal safety awareness and invited me to quote for training his employees in behavioural safety. When I visited their facility, I noticed that there were several machinery safety guards missing and lifting equipment had been illegally modified. I declined to quote for the work on the basis that they had got the balance wrong. There was no point in further refining their people skills if the most basic equipment safety standards were not being enforced. If you have good people but poor equipment standards or out-of-date procedures, you will still have accidents. The three-legged stool concept (see Fig 7.4) is intended to convey the message that if you are paying less attention to one of the 3 "P"s, then the safety stool will be unstable and it will fall over. To dramatise this concept and make it more memorable, I have made a model of a three-legged stool (Fig 7.4). Having explained the idea, I place an image of a person on the stool. I indicate that if all the legs (i.e. the 3 "P"s) are equal, the stool is stable and the person is safe. One of the legs in the model has a detachable section to reduce its length, and so I demonstrate that if one of the legs (i.e. the 3 "P"s) is shorter than the rest the stool becomes unstable and the person's image crashes dramatically to the floor. It takes 15 seconds to show, but the image and therefore the concept stays with the trainees for a very long time.

Figure 7.4 The three-legged stool.

If the idea of three"P"s doesn't suit your situation, other people use the TOPS acronym for the same message, where TOPS is:

- Technology
- Operations
- People
- Systems

The only problem using the TOPS acronym is that a four-legged stool doesn't tend to fall over so easily and so the model message is lost!

7.4 POSTERS AND CHART BOARDS

Tutors should use every opportunity to reinforce important messages. Posters and charts not only brighten up a training room or workplace, but they provide an opportunity for participants to assimilate information even before a training session starts. Having relevant poster information around the walls gives people something to look at and question even while they are waiting for the session to start, or during coffee and lunch breaks. There are endless sources of safety awareness posters available in the marketplace or on the internet, many of which are well produced and have quite high impact. Just make sure that the ones that you use are relevant to the subject that you are covering (Fig 7.5).

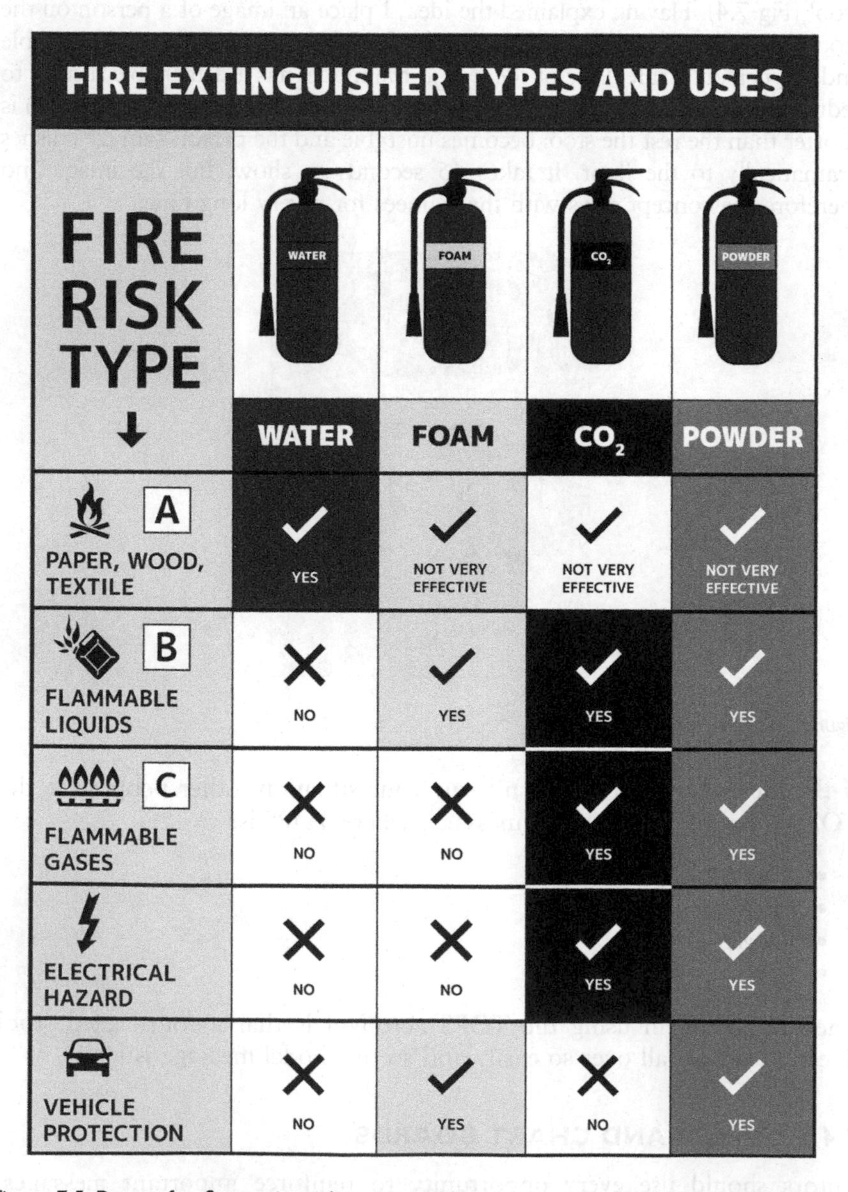

FIRE EXTINGUISHER TYPES AND USES

FIRE RISK TYPE	WATER	FOAM	CO_2	POWDER
A PAPER, WOOD, TEXTILE	✓ YES	✓ NOT VERY EFFECTIVE	✓ NOT VERY EFFECTIVE	✓ NOT VERY EFFECTIVE
B FLAMMABLE LIQUIDS	✗ NO	✓ YES	✓ YES	✓ YES
C FLAMMABLE GASES	✗ NO	✗ NO	✓ YES	✓ YES
ELECTRICAL HAZARD	✗ NO	✗ NO	✓ YES	✓ YES
VEHICLE PROTECTION	✗ NO	✓ YES	✗ NO	✓ YES

Figure 7.5 Poster for fire extinguisher recognition.

When using posters, do make sure that they are changed regularly. An ageing and yellowing poster will convey the impression that you are not serious about safety. The best and most economical way of keeping poster campaigns and information fresh is to rotate your posters through a series of different locations. This way, each area sees new posters every week or two, but you are doing it in a way that does not compromise your budget. Some poster suppliers also sell easy fit frames, so that you can easily change posters without them becoming torn or dog-eared.

In this day of desktop publishing, it is quite easy to produce your own posters with bespoke messages containing relevant photographs of your workplace. Ideally these should be produced at a size larger than foolscap/A4 to have the desired impact. By delegating the production of such posters to a small team from the shop floor, you can get greater involvement and ownership of the messages and ensure that safety is everyone's responsibility and not just a management diktat.

When running training sessions that have a series of linked components, it is a good idea to have a large flip chart-sized foam board with the key steps that you are trying to cover. As you progress through your presentation and exercises, keep referring back to the board to show which step you are at. This not only gives the trainees an idea of how much more there is to go (they are mainly interested in when you will finish!) but because the board is always on display, it continually reinforces the system that you are following and hence will be held in their long-term memory (Fig 7.6).

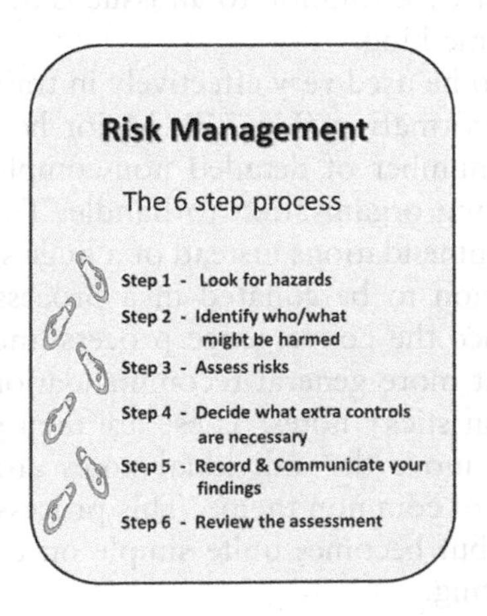

Risk Management
The 6 step process
Step 1 - Look for hazards
Step 2 - Identify who/what might be harmed
Step 3 - Assess risks
Step 4 - Decide what extra controls are necessary
Step 5 - Record & Communicate your findings
Step 6 - Review the assessment

Figure 7.6 Example of a foam board poster for use on a flip chart stand.

Home-made posters can also be valuable when you are mentioning a subject that you expect most people will be familiar with. If there is anyone who is not familiar with it, you can refer them to the details on the poster and say that you will explain it to them in detail at the end of the session. This means that you can avoid boring those who already know and avoid losing time in your training session.

I find that many safety managers office window-cills are festooned with relics of bye-gone incidents. Photographs of these relics can form the basis of some very useful poster campaigns.

7.5 STICKY NOTE DISPLAYS

Self-adhesive "Sticky Notes" can be used to get participants involved in training. Everyone can have access to a sticky note pad and instead of the tutor asking for feedback or input to the group at large, getting everyone to write down their ideas on a sticky note, ensures that it is not just the person with the largest ego who responds. When asking questions, phrase them is such a way that responses are concise and participants do not have to write an essay in response. Once everyone has produced their sticky notes, invite them to stick them onto a wall or board and ask them to group them in common ideas. This approach not only gets everyone's ideas considered, but also gets them moving around and working in ad hoc groups while they discuss what notes can be grouped together. This can be a very powerful way of getting a group to have ownership of some ideas, because they will see that they produced the ideas and were not just told the answer by the tutor. Developing your own solution to an issue is the most likely was that you will remember something.

Sticky notes can also be used very effectively in training people to collate apparently random information (Fig 7.7). Major health and safety audits can generate a large number of detailed non-compliances which may be difficult for the receiving organisation to handle. To provide broad management systems recommendations instead of a huge shopping list of detail, requires this information to be collated in a process called convergence. Trainees can experience the convergence process and how details can be merged into larger but more general recommendations by putting the detailed audit findings on sticky notes. These are then placed on a board or wall and the trainees move the individual notes around to form groups which have some sort of common theme. This process may sound complex to explain in writing but becomes quite simple once demonstrated by the tutor in a training setting.

Figure 7.7 Using "Sticky Notes" to rationalise information during a health and safety audit.

7.6 REAL-LIFE EXPERIENCE

Some of the best tutors are also good raconteurs. A raconteur is someone who excels in telling anecdotes. In a training environment, a good raconteur tells stories that are of interest and entertaining to the participants and which complement and develop the topic being discussed. To be an effective raconteur needs experience in the topic but also a flair for holding people's attention. Telling relevant stories that reinforce a point made in the training will provide high impact, confirm what is being said is not some hypothetical invention of the tutor and will embed the point in the listener's "long-term memory". The danger even for good raconteurs is related to time management. It is always rewarding for a trainer to get lots of questions, but well-told and relevant anecdotes will undoubtedly lead to lots of questions. This shows that the raconteur has captured the attention of the audience. The problem is knowing when to stop. Even the best raconteurs often have difficulty controlling time. It is important to remember that during each training session the tutor should have identified a clear "route map" on which he or she will have identified a series of key steps, tasks or pieces of information that need to be shared. Once time starts to over-run, then there is often a tendency to rush the latter parts of the training and potentially miss out on one or more of the key steps.

When explaining real-life incidents, it is important to select examples that will stimulate interest and discussion. Some of the big international disasters or even the more significant local incidents are not necessarily the best examples, because many people will have heard about them and studied the learning from them before. In this case, the trainees are not applying much brain-power to the event because they already know all the answers. In selecting the relevant anecdotes, the training session designer must be aware of what has been said before in relation to the examples that he or she intends to use.

I was training a multi-disciplined group from several different countries and industries in "Safe Systems of Work". To select anecdotes and exercises from one of the industries present would have given some people who had the right technical knowledge advantage and would have left others bemused. In this situation I decided to use examples which were from none of the industries represented, but to which everyone could relate. I chose examples of accidents from the aviation and railway industries. Everyone understood the events that were described and could see the learning without feeling in some way responsible or guarded about their responses. Examples from these two industries are immensely valuable to the tutor because they are some of the safest industries in the world, and the aviation industry has arguably the best incident investigation and reporting process in existence. There is also a wealth of information available, which tells about travel-related accidents and what learning arose from them.

The most important message about telling anecdotes is to ensure that they are relevant and that trainees will have to think about them, rather than just trot off the answers from memory. It is important for the effective tutor to develop a portfolio of stories that help explain the topic. These might be "excellence" stories which emphasise best practice, "horror" stories that emphasise bad practice or "incident" stories which allow the trainees to learn from other people's mistakes. Particularly in the case of the excellence or horror stories, be prepared to draw the parallels with the situation that exists for the trainees. When I was training a group of managers from a chemical company in safe systems of work, I use a tragic accident example from the rail industry. One might ask the question "what has a rail accident got to do with chemicals processing?", but in terms of the safety systems that were being used they are directly comparable. Not only that, I find that if you use examples that everyone is familiar with, then they probably know all the answers to the obvious questions without having to think about it. If you use anecdotes or examples that they are not familiar with, then they really have to think about it. Using examples from outside the trainees' immediate experience also means that you avoid embarrassing anyone who might have had an involvement in the story being told.

7.7 SIMULATIONS (SEE ALSO RESOURCES SECTIONS K AND O)

One of the most powerful means of learning is the practical "hands-on" approach. However, there are some potential problems with just going out onto the shop floor and doing things. The first issue is that particular learning situations can occur randomly, and it may be that the set of circumstances that you wish the trainee to experience is just not happening on the day of the training session. More importantly, as we are talking about health and safety training, it may be that we are wanting to learn about a particular hazard and it is possible that hazard could lead to an uncontrolled risk of injury. We have already learned that in the present day, more western military personnel are killed in training exercises than are killed in combat. We therefore need to be very careful that our training does not expose the trainees to unacceptable risks.

One of the best ways of controlling the risk associated with trainees getting practical experience is to set up a carefully controlled simulation. It may be that you want to teach maintenance technicians about what constitutes a safe scaffold. Rather than taking them into a congested normal working area, it might be possible to set up a scaffold simulation in a safe area where there is no vehicular traffic or risk from daily production activities. The other benefit of this approach is that the tutor will have identified the learning points in advance and so common errors in the scaffold design could be built into the scaffold so that the trainees can identify them for themselves. By carrying out a risk assessment of the simulation, this can be done in a way that minimises risk. Furthermore, by photographing and videoing the simulation, the set up can be used and reused repeatedly in future training with no risk of injury at all. Fig 7.8 shows a scaffold tower that has been built with a series of design faults for trainees to identify. The simulation was then photographed and videoed. The photographs and videos were then used for training purposes, rather than the faulty scaffold itself.

Figure 7.8 Scaffold tower demonstrating design and construction faults.

Simulation involving learning on chemical processes can be particularly valuable in order to avoid the risk of trainees being exposed to chemical hazards. This can sometimes be done using obsolete, redundant or out-of-service equipment. However, the best simulation I have done was whilst teaching safe systems of work to a group of international chemical industry managers in a rather plush hotel. We were nowhere near an operating factory and in any case, time was tight and would not have allowed for transporting delegates to an industrial facility. Instead, we discussed the matter with the hotel management and got their agreement to use the hotel's biomass boiler house in a simulation. We needed to get small groups of delegates to carry out a syndicate exercise that involved preparing a piece of chemical plant for maintenance to deal with a leak of flammable liquid. The hotel's boiler system was of course in reality processing water, but we relabelled the pipes, pumps and tanks with temporary hazchem labels and valve and equipment numbers to simulate a situation that we told the delegates was processing hot flammable oil. There was an oily spillmat placed on the ground in the location of the leak and various identification tags were placed around the pipework. Piping and instrument drawings were provided based on modifications to the hotel's own drawings. The benefit of this approach was that the delegates were able to do a "hands-on" study of the "hazardous process" in a very low-risk environment (more information on this example is in the Resources Section O) (Fig 7.9).

Figure 7.9 How a safe water system can be simulated as a high-hazard system.

Kinetic handling and manual handling training are areas where during the session, practical work might lead to injury or musculoskeletal strain. To avoid this happening, you can simulate the loads involved. If the loads are boxes of product, then the manual handling can be practised by using part empty boxes, or if bags of product need to be handled by hand, fill a product bag with plastic packing or empty plastic drinks bottles. Don't make the load so light that it can be lifted with one finger as the trainees will not take the exercise seriously.

Physical simulation training can also be done quite effectively prior to the commissioning of new plant or equipment. Even if the equipment cannot be operated, letting operators see, touch and comment on new equipment before it is used, can be a very helpful learning experience and can often identify potential future problems that the designers did not foresee. This technique is known as "Pre-Start-up" Simulation.

The only issue with simulations is that they do take time to design and set up.

7.8 ROLE PLAYS (SEE ALSO RESOURCES SECTION I)

Role plays are a fairly well-established form of training. Role playing is not about testing trainees' thespian abilities. It is about re-creating reality. The role play will be based on a case study which may be an actual event or a fictitious event created for the purpose of learning. As with so many exercises, the success of a role play is largely down to good preparation and planning. It also requires the tutor to have some knowledge of the trainees in advance. Role play usually requires some extrovert trainees in attendance. The role play is normally used to allow trainees to practise some knowledge or technique that they have learned about earlier in the training. It works particularly well for training people at all levels to become comfortable with safety communications and discussions. Very often this will be in a situation where an organisation is introducing a behavioural safety or safety observation programme or health and safety auditing training.

In order to maximise the learning for all trainees, whenever a role play exercise is carried out, it will need to be observed by the other trainees and also the tutor. On completion of the role play, the observers should be invited to comment on what went well and what did not go so well. The tutor then provides detailed feedback.

To allow the role play to be successful, it is essential that the tutor provides a written brief for each of the players. If the situation portrayed is at all complex, it is helpful to discuss this with the players well in advance. To create a really high standard of role play, consider video filming the scenario that you wish the players to discuss. This is particularly useful and realistic when training people to take part in behavioural safety discussions, as you can video their own workplace in advance and then get them to discuss whether there were any unsafe acts or conditions and what should be done about those.

In order to get players into their role, it is a good idea to give them some props or indication of the fact that they are playing a role and not necessarily being themselves. I would normally suggest that the players might wear a Hi-Vis vest or hard hat to help get them into their character.

When providing feedback on the role play, remember that you are not commenting on their acting ability, but only about whether they are adopting the training and skills that you have previously taught them. It is also important that the feedback is not seen to be critical. Always start off by praising them for what they did well and how they set the other person at ease. If you observe that they haven't followed the process that you advised, talk about what they did and then ask:

"How do you feel that the other person felt about that?"

or alternatively

"How do you think you could have done that differently?"

Role plays are usually fun and enjoyed by both participants and observers. If you are training a large group at the same time and you want everyone to take a turn at the role play, this can take a lot of time, but also after a few attempts, it tends to be repetitive and little new learning is derived. Role plays are best utilised in small groups, unless two tutors do the role play in front of everyone else in order to make a particular point. To do role plays well takes preparation, planning and time. When using role play, it is important to ensure success that:

i. There is a clear purpose to role play
ii. The role play supports the topic being taught
iii. It is realistic
iv. The trainees can identify with the subject
v. It is not a threat to introverted trainees

7.9 PHOTOGRAPHS (SEE ALSO RESOURCES SECTIONS B AND U)

Using photographic images is one of the best and easiest ways of creating impact and gaining understanding. Photographs can be used very effectively in work instructions, training manuals, posters, presentations and exercises.

7.9.1 Photographs in Work Instructions and Training Manuals

Quite often I find that work instructions are not very "user-friendly". Work instructions are primarily for the use of workers at the shop floor level, but they are invariably written by managers. This means that they often contain reams and reams of words. A particular example that I came across recently was over 80 pages of "legal-speak". It is hardly surprising that workers do not really read them. Instructions should be clear and concise with an indication that if there is a need to diverge from the defined work method, there is a simple process for referring up the management structure. It has always been said that "a picture is worth a thousand words" and that is true when it comes to instructions and training manuals. It is no accident that home D.I.Y. enthusiasts find that when they come to assemble their latest delivery of flat-pack furniture, that the instructions are in the form of pictograms. Not only does it make instructions clear to follow, but it also deals with the increasingly common situation, that not all workers necessarily can read the local language.

I have found that in the case of relatively straightforward tasks the best form of instruction is to have a laminated version of the pictorial instruction posted immediately in sight of where the work is normally carried out. If the task is to be done step-wise then there should be a photograph of each step taken from the most relevant angle. The great thing about this is that the photos can be taken by people who actually do the job. Good quality photographic images convey so much information to the worker (Fig 7.10).

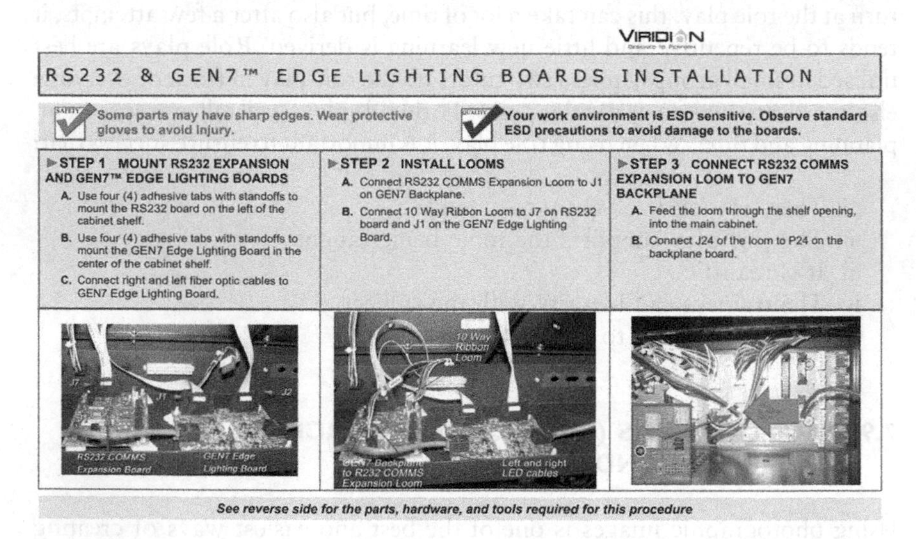

Figure 7.10 Example of a pictorial instruction courtesy of Mary Garnett (McDaniel).

Maintaining housekeeping standards is often an ongoing challenge in some facilities. When the desired standard of housekeeping has been achieved, take a photograph of the area and post it on the wall, so that everyone knows what standard is expected and the photograph is there to prove what is possible.

Photographs are equally valuable when it comes to maintenance manuals where it is possible to use photographs to define the sequence of disassembly or re-assembly. They can also be used to specify the standards of cleanliness required and to demonstrate typical faults.

7.9.2 Photographs as Posters

Posters are a good way of reminding people of important behaviours or about some new learning that arose from a recent incident. However, poster information must remain fresh and relevant and be posted in places where they can be both seen and read. For self-produced posters I would recommend that photographs should take around 50% of the space, the remainder being explanatory text. Remember that it is not the objective of the poster to embarrass individuals or show accident victims. If you think that there is learning in showing the consequences of someone's injury, then that can only be done with their approval (Fig 7.11).

Figure 7.11 Self-produced safety awareness poster.

Many organisations have a "safety alert" procedure whereby they notify all their employees about learning events or injuries through a simple computer-printed poster containing photographs of the location of the event and the immediate learning points in bullet form (Fig 7.12).

SAFETY ALERT

FATALITIES ON AUTOPACKING LINES

There have been reports of two fatalities on auto-packing lines with identical guarding design to this plant. In both cases the operator entered the danger zone without using the interlocked access gates. One of the cases occurred at a petfood manufacturing plant. The incident happened during the nightshift. It is believed that the operator noticed a damaged bag on the palletiser and then entered the danger zone by stepping from the palletiser upper platform and stepping over the handrail at the top of the stairway. This allowed him to go around the end of the fence guard and enter the feed level of the palletiser, by-passing the interlock system.

At this time the palletiser was not in operation, but it was not isolated. The operator found his way to the ground level and started work to clear up the small spillage that had occurred. It is believed that at some stage he must have inadvertently contacted a limit switch, because the pressure plate was activated and he was fatally crushed.

You are reminded of the potential dangers of entering the danger zone with the power not isolated. Statistics show that if you are injured inside the danger zone when the power is still on, it has a 1 in 4 chance of being fatal.

Remember that on auto-packing lines there is no such safe condition as "STOPPED". Auto-packers only have two stationary states. They are either:

"ISOLATED", or "JUST ABOUT TO START"

Figure 7.12 Safety alert.

7.9.3 Photographs in Safety Training Presentations

We have mentioned previously in Chapter 2, the importance of using images to aid the memory. Photographs are probably the easiest, clearest and have the highest impact of any of the images that can be used in training. Photographs that are embedded in a PowerPoint presentation can serve several different purposes. They can be used to:

- Clarify understanding of a situation that you are about to discuss
- Identify a hazard
- Be used interactively to get the audience to identify what is wrong in the photograph (i.e. get them talking and engaged)
- As a light-hearted interlude in proceedings

I have a principle that "I can teach anybody anything if they are smiling" and so I am a great believer in humour to help make training memorable. This is a fine line to tread. Safety is a serious subject and so you must choose your moment carefully, but the type of humorous photographs that show health and safety gaffes are readily available on the internet. Provided they are not offensive to any attendees, these types of photographs can set people at ease and provide a bit of fun in the day. The only problem is that most of the internet-sourced photographs are so extreme that the trainees could reasonably argue that they would never do anything like that. If you get that sort of response then I just ask people whether the lack of the right equipment has ever stopped anybody from doing that "essential" job at home. This also gives you the chance to remind people that being at work is probably the safest time of the day for most people! (Fig 7.13).

Figure 7.13 You can teach anyone anything if they are smiling!.

7.9.4 Photographs Used in Training Exercises

Every exercise or syndicate session needs careful planning in order to ensure that it is relevant to the subject being taught, interesting for the trainees, has a tightly controlled timescale and gives the opportunity for all those involved to participate and give feedback. Time is nearly always at a premium and so there is not usually an opportunity for very lengthy explanations – the tutor should have available a very clear and concise brief for each group, which should be easily understood. It is often useful to include photographs in the exercise as these can give a wealth of information for the group to analyse (Fig 7.14). When working in a training centre setting (i.e.

Risk Assessment Exercise

Syndicate Exercise – Group 1

Consider the situation shown in the photograph above. The scaffold is to provide access for painting.

Answer on a flipchart:

1. **Why has the scaffold been designed this way?**
2. **What could go wrong?**
3. **What are the possible consequences?**

Be prepared for one of your syndicate group to provide feedback to the rest of the course.

Figure 7.14 Exercise for understanding risk and consequences.

away from the workplace) with the availability of small "break-out" or syndicate rooms, then photographic information is particularly useful in creating realism and relevance in exercises involving:

- Hazard recognition or hazard spotting
- Risk assessment
- Audit training
- Housekeeping awareness
- Chemical hazard labelling
- Machinery guarding
- Behavioural safety actions

The great thing about using photographs in syndicate exercises is that the trainees will invariably come up with points that you, as the tutor had not expected. By doing this the participants will feel good that they have identified something that no one else had thought of!

7.9.5 Photo Hazard Spotting Exercises (See Also Resources Section B)

One of the most common feedback complaints about a training course is that it was boring. It is important to remember that many practical workers are not used to having to sit in one place for an extended period. You should take every opportunity to get people to move around and to do things for themselves, rather than just being lectured at! One way of getting involvement is to use a photo-spotting exercise. This entails preparing a set of 20 or so photographs which demonstrate examples of particular points that you want to make – it may be examples of occupational health hazards, physical hazards (Fig 7.15) or even ergonomic hazards. Each photograph is clearly numbered (1–20) and these are randomly located around the walls of the training room. It can also add to the interest if some photos are placed in other easily accessible places like the corridor outside the training room. It is important that the photographs are not so close to each other that people are bumping their shoulders to look at them. I usually break the attendees into groups of two, so that as they are going around there is some discussion amongst the pairs. Each pair is given a pre-prepared answer sheet which has an un-numbered list of 20 descriptions of the photographic hazards and they are asked to decide which photograph represents which hazard. It takes about 20 minutes to complete including the feedback. More details of this type of exercise are shown in the Resources Section B at the back of this book.

Figure 7.15 Example of photo spotting exercise.

7.9.6 Photographic Hazards Spotting (Camera Hunt See Resources Section C)

One of the most powerful ways of getting involvement in training is to get the trainees to take the photographs for themselves. I use this type of exercise primarily for educating people in hazard recognition and housekeeping standards. It is most appropriate when the training is done by employees in or close to their own workplace. Very often people get to accept hazards and unsafe conditions that they see in their daily work, either because it has not caused them a problem yet, or because things have changed incrementally over a period of time, so that between one day and the next, there is no obvious change. It is only when people stand back and see how things have evolved over a long period of time that they appreciate the significance.

The Camera Hunt exercise needs to form a part of a structured training session, during which time the trainees are given understanding of what sorts of hazards might exist in their workplace. Having completed that part of the training, the group is divided into small teams of three or four and each team is given a camera and allocated a workplace area to go and identify and photograph hazards. It is a good idea for some of the trainees in each team to be from the area to be studied, so that after the training is over, they can actually implement some corrective actions. For this exercise to be effective, takes time and usually I would allow at least an hour for this. Just be careful about selecting the areas for study, as there may be somewhere it is not possible because of Atex classification (de-matching), business confidentiality or other hazards.

You will be astonished at how many photographs will be taken. It is not unusual to have over a hundred examples of hazards identified during the hour in the workplace. This provides a wealth of information for both workers and management and it is important that this valuable information is not lost. On returning to the training room, each team should be invited to display and comment on their photographs, identifying the hazard and what they think should be done to control that risk. This then reinforces the learning in the minds of the team but also shares that learning with the other teams. You need plenty of time to share this feedback if there are several teams, but remember that this is a key part of the learning and it should not be skipped over (Fig 7.16).

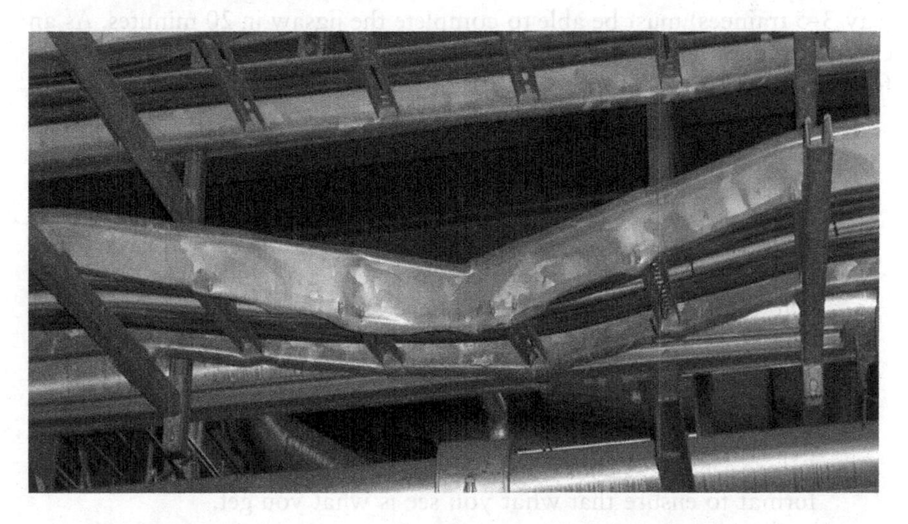

Figure 7.16 Example of photo spotting hazards – forklift truck impact.

Every training event should not just be a snapshot in time. There should always be a longer-term consequence of the training. One way of extending the impact and longevity of the training is to use some of the photographs to create posters around the workplace. I would usually make the final stage of the hazard training an opportunity to let each team select one or two of their most significant photographs and create a hazard awareness poster for their particular workplace. It can be helpful for the tutor to pre-prepare a standard format for the poster, so that all the teams need to do is to copy and paste their photograph and type in a short learning message.

In order to make this of on-going benefit, I then encourage each team to take on a role of using other photographs that they have taken to produce a new poster each month.

7.10 JIGSAWS (SEE ALSO RESOURCES SECTION A)

A fun way of conveying a complex concept is to start by using a jigsaw. Everyone knows how to do a jigsaw, and so they can be very good ice-breakers for trainees who may not know one another. As groups work to complete the jigsaw, they subconsciously start to take in the details of the jigsaw image and reinforce memory. These can be particularly helpful in conveying messages that involve block diagrams and are best used just before going into a detailed explanation.

There are endless suppliers of bespoke jigsaws of varying quality available on the internet. I tend to use an A4 size, because those are readily available, but in general terms the larger the size the better. The main problem with jigsaws is that once started you really have to finish it! The snag is that it is very easy to make a jigsaw impossible to do, or for it to take a disproportionate amount of time. To have any real value, a small team (say 3–5 trainees) must be able to complete the jigsaw in 20 minutes. As an encouragement and for a bit of fun I always offer a prize (chocolate bars) for the first team to complete their jigsaw.

When designing a jigsaw for the first time, there are a few rules:

i. Do not use jigsaws of more than 80 pieces – they take too long to do.
ii. Areas of plain colour which are larger than the individual jigsaw piece can make the jigsaw impossible to complete. Always use a background which is patterned or which has a faded photograph to help guide where the individual pieces go.
iii. The key information should stand out clearly. To aid learning use reasonable-sized lettering in a white background box, with coloured edges.
iv. Do not use large amounts of text.
v. Before sending the jigsaw image to the printers ensure it is in pdf file format to ensure that what you see is what you get.

Once the jigsaws have been manufactured, ask a colleague to complete one and time how long it takes. Ensure that this can be done within the time-scale that you have available (Fig 7.17).

When preparing to use the jigsaw, lay the pieces out picture side up to avoid losing time for the team just to turn them over. At the end of the training session check the number of pieces before putting the jigsaws away for use next time – there is nothing worse than getting near the end of a jigsaw to find that there is a piece missing.

Once the jigsaw is complete, remind the teams to "read the message on the jigsaw" as this is part of the learning process and not just a fun break! Then make sure that when returning to your presentation, that the next slide replicates the message from the jigsaw exercise, so that it can then be seen in context and explained.

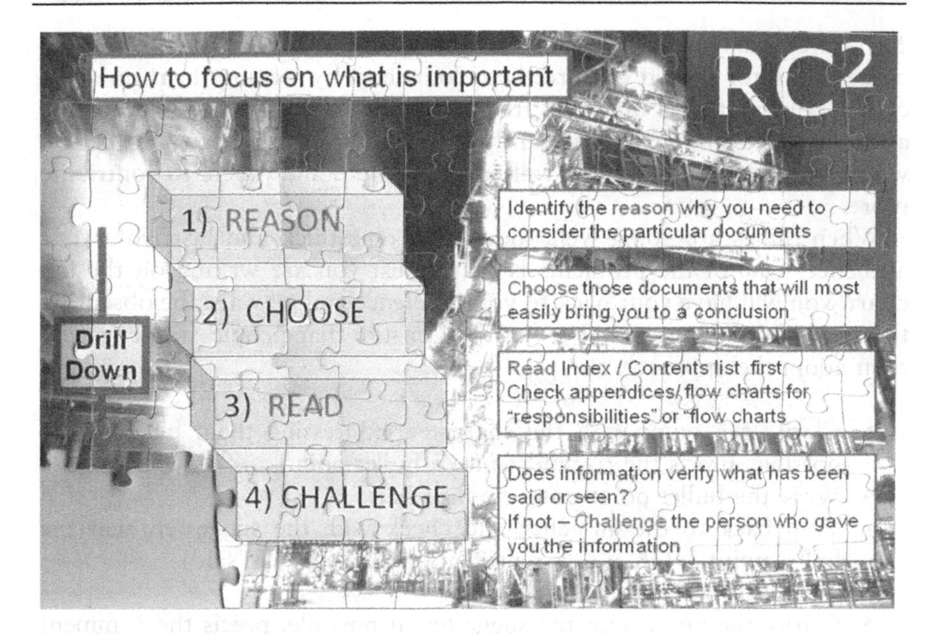

Figure 7.17 Example of safety training jigsaw.

7.11 USING FLIP CHARTS

Flip charts are an essential, cheap and a readily available tool in the trainer's toolkit. They are very valuable in helping to engage trainees and make the session interactive. Unfortunately, in this digital age, trainers are using flip charts less and less.

In my opinion, the success of a training event is largely measured by the amount of interaction that goes on between the trainees and the tutor. If there is some good-hearted banter, then things are going well and the trainees are engaged and will have the best chance of understanding the messages. On the other hand, if there is no interaction, then people are either bored or they are not understanding what you are saying. In this situation things are not going well.

Flip charts can be used for various purposes:

a. To help engage the trainees
b. Capture relevant feedback
c. To show that you are listening to what trainees are saying
d. To describe the circumstances relating to an anecdote that you are telling
e. To explain an answer more clearly to a question posed by a trainee

Flip charts are also the standard method of requesting formalised feedback from a syndicate exercise or break-out activity.

In order to engage trainees, they need to be encouraged to participate. To make things interactive, the tutor will need to pose questions to the audience.

These are usually referred to as "plenary questions". When good responses come, the tutor should capture these comments by writing them on a flip chart. This reinforces messages that you have either already made or are about to make, but more importantly it gives some recognition to the person who gave the response and it will encourage him and others to contribute more.

When things are going well, ideas will come thick and fast from your audience. One of the problems is that whilst you are writing on the flip chart, you will have your back to your audience and may also be obscuring the chart from some people's view. To ensure that people stay engaged, then adopt the approach of:

- Clarify the point with the suggester, addressing them by name (if possible, précis the comment into a bullet point)
- Write the bullet point onto the flip chart
- Turn back to the audience and check with the suggester, that the bullet point is OK (i.e. it represents what he or she said)
- Talk and receive other suggestions
- Clarify the point with the suggester (if possible, précis the comment into a bullet)
- Write the bullet point on the chart
  Repeat the process as long as is necessary.

The important thing is in order to engage with people, you should not have your back to them too long. Being able to capture feedback quickly and legibly comes with practice. One technique to avoid turning your back on the audience is to invite one of the trainees to come up to the chart and write up the ideas for you.

If your interactive session is aimed at drawing out some particular points, then a good tip is to write these points lightly in advance with small pencil lettering on the flip chart. As you are right next to the chart, you will be able to read your prompts, but other people who are further away will not be able to see them. This is particularly useful if you are trying to get trainees to develop a sequence of actions, but their responses may come in a rather random way. This method means that the trainees appear to come up with the ideas but you are defining the sequence or filling in gaps. Information and concepts arrived at through this method will have a high level of ownership from the trainees and therefore they are more likely to remember it and to apply it later.

I often find that some really valuable learning can be captured on the flip charts and it is very motivating during the session to refer back to points which individuals made earlier. The problem with flip charts is that each time a new one is used, it tends to cover up the previous chart. If you know that a comment may be relevant later, or even if you want just to encourage involvement, I find it useful to remove completed charts and stick them on a wall or pinboard. If you are meeting in a hotel or the like, then just check that the hotel management are happy with you using Blu-Tak or draughting tape to stick things on their walls! They tend not to appreciate having to repaint the walls after each use. Many purpose-designed training rooms have magnetic metal wall strips or clips to allow for the posting of completed flip charts. Alternatively, if you leave the charts on the pad, but know you will need to come back to a particular chart later, then a tip is to turn over the bottom corner of that particular chart or leave a Post-it note sticking out so that you can find it again quickly.

Before starting your session, ensure that you have some spare flip chart pens that will write. Do not use green as it cannot be seen at a distance. Black, red and blue are best. Just remember that if you have flip charts on the wall and want to add points to an existing chart, never use a permanent marker as this will go through the paper and ghost write onto the wall!

If you want to avoid ripping your completed chart across the middle when tearing it off, another tip is to carefully tear a starter tear of a few inches of each chart in advance. I also find it best to remove the stubs of previous old charts from the pad before the start of the session as poor previous tears tend to perpetuate themselves!

The last point about flip charts is about the positioning of the flip chart stand. Before you start, try sitting in the trainees' seats and ensuring that they will all be able to see the charts and that the chart stand does not obscure part of the display screen. The most common problem with flip chart stands is that because the legs tend to stick out, they are also a frequent trip hazard for the tutor! It tends to set a bad example in safety training if the tutor trips over the flip chart stand!

If the tutor is using a combination of a PowerPoint presentation and flip charts, I find that it is useful for the presenter to have a discrete prompt on his or her slides to indicate when to invite an interactive discussion. The reason that this is important is that any handouts should have been prepared so that the information relating to this particular discussion is hidden, otherwise trainees will just read back the answer without thinking. There are several ways of providing an effective prompt. My favourite method is to use something that is on the background of every slide. It may be a line under the slide heading or some aspect of a corporate logo. By changing a colour in the line or logo lettering or placing a coloured line at the base of the slide, the tutor will see that he has to seek contributions from the group and will know that the answer is not printed in the session handouts. The trainees will not understand the significance of this and probably will not even notice the minor amendment in the presentation slide.

7.12 DRAWINGS AND WALLCHARTS

Every opportunity should be taken during training to use different ways of getting training messages across. Some organisations that have drawing/design offices may already have poster-size printing facilities. If not, these can be done quite economically, either in some of the larger office equipment stores or from internet-based suppliers. Poster information can be used during periods in the training when tutor-led training is not happening. In particular it is useful to have some interesting and relevant information available on the walls or display boards for that unproductive time before the training starts and you are

What is an Occupational Exposure Limit?

A definition:

"The concentration of a substance in air, averaged over a specified time reference period ,representing conditions under which it is believed that nearly all workers (assuming normal healthy adults) may be repeatedly exposed, day after day, over a working lifetime without experiencing adverse health effects."

Full Shift Exposure assessment (Usually taken to be 8 hours)

Sum (task assessments + fugitive emissions) = shift average exposure

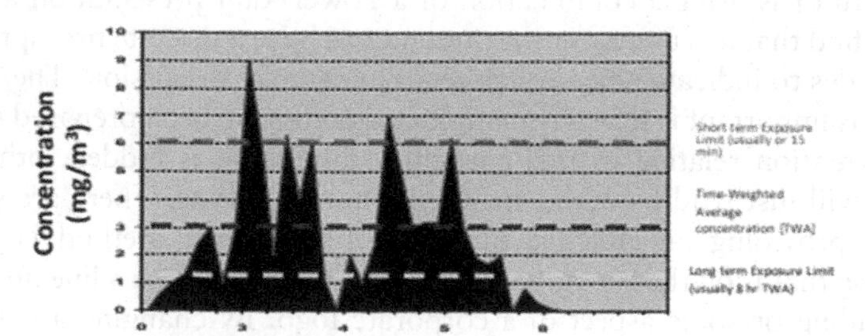

- Personal monitoring may be needed to estimate extent of full shift exposure

An OEL is a risk management tool to indicate the standard of control required to ensure good health is maintained

Figure 7.18 Example of self-produced health poster.

waiting for the full complement of trainees to arrive, or during breaks in the structured training process. If the training is taking place in a conference centre or location where you cannot put up posters on a permanent basis, then I keep a set of cheap light cork pin boards that I can attach the poster to in advance and just hang the pin board over wall pictures or stand them on tables for the duration of the training (Fig 7.18).

7.13 PROPS AND GIZMOS (SEE ALSO RESOURCES SECTION F)

If the training involves the use of some specific equipment, then it is important that the equipment is available at the time and can be safely handled by the trainee. A classic health and safety training example of this would be when employees undergo fire safety and extinguisher training. Although it is appropriate that people know the basic theory of the fire triangle, the overriding purpose of fire extinguisher training is to know how to use a fire extinguisher. No amount of theoretical training will prepare someone for the noise experienced when a CO_2 extinguisher is activated or how it feels to be close to the heat of a fire. In many cases health and safety training can only be effective with some "hands-on" experience of specialised tools or equipment. When carrying out training practical work which might be potentially hazardous, always ensure that it is done by a competent trainer and that the training exercise has been subjected to a satisfactory risk assessment (Fig 7.19).

Figure 7.19 Fire extinguisher training.

This is also the case with Personal Protective Equipment (PPE) training. Many organisations almost automatically resort to using PPE as the default method of risk control without the specifying manager really understanding what it is like to wear PPE for a long period of time. Many years ago, one of my managers wanted to make just this point. We were asked to turn up for a training meeting dressed in full PPE. We had to wear a full PVC chemical suit, hard hat, goggles, ear defenders rubber gauntlets and rubber boots. He then told us to wear that equipment for 4 hours, during which time we were put through a series of physical tasks. It very effectively brought home to us that not only could we not hear what we were being asked to do, but our goggles quickly steamed up, limiting our vision, the rubber gloves allowed no digital dexterity and so we couldn't write or operate small tools like electricians' screwdrivers and finally within a few minutes of doing quite mild exercise we were sweating profusely within the impervious chemical suits. Quite soon we were at risk of getting heat exhaustion and our cognitive ability was impaired. This was a really effective way of learning and is an experience that I have never forgotten. The key message here is be sure you understand what are the adverse consequences of what you are asking your subordinates to do? Remember that like any prescription medication, every remedy has an unintended side effect.

If the training involves the correct way to wear or remove protective equipment, then do not just tell people how to do it, but show them and get them to practice it themselves – that way they are more likely to remember (Fig 7.20).

Figure 7.20 Getting managers to understand what it is like to wear PPE for long periods of time.

Likewise, if the training involves the use of tools or equipment, such as a noise meter for sound measurement or a gas analyser for confined space entry training, the minimum requirement is that the trainees actually get to use the instrument themselves, even if the training is just at the awareness level. For example, for the confined space worker, being aware of what the competent or licensed atmosphere testing technician should be doing is an important cross-check in the confined space entry approval system.

In addition to using actual tools and equipment, there are other hardware training aides that can help raise interest levels and make the training more memorable. I refer to these as training gizmos. Everyday working involves assessing and managing risk. Those who are familiar with behavioural safety concepts will already know that what motivates us to behave either safely or unsafely are the perceived consequences of our actions. If we consider that there is a benefit to us to take a shortcut, then we have an incentive to do it. On the other hand, if we think that there is a serious penalty to our planned action, i.e. we get injured or might get dismissed, then we might be less likely to take this action. On balance if we think the risk is more likely to give us a benefit than a penalty, then we are motivated to do it. Managing this type of risk decision is the golden thread running through all health and safety training. It is important to get trainees to understand this risk/consequence idea in a way that they remember it and are not harmed by the learning process.

To this end I have produced for myself some little gizmos that very effectively convey this message. The first one is based on a "drinker's game" that I found in a local shop some time ago. This is a little random generator that by pressing a button tells the six players which one has to take a drink next. Although I have never tried it, I gather that the winner is the player left standing when all the others have slid under the table!

I have modified the game by producing an overlay and gluing to the top. Now the pressing of the central button is an indication of the player taking a chance. At each of the six random indicator lights is a consequence. Some of the consequences are positive (the player wins a chocolate bar) and for others there is a negative consequence, in that the player gets injured (and has to wear an arm sling) or they have an effect on someone else. There is also one outcome where nothing happens (equivalent to a near-miss) (Fig 7.21).

The game quite accurately reflects real working behaviour. It is much more common that trainees take the risk and get away with it (i.e. win chocolate!) than are injured (i.e. wear an arm sling). This shows the trainees that we often get away with taking chances, and that encourages us to do it again. The game also shows us that taking chances might affect other people rather than ourselves. The interesting thing is that although I do not make it compulsory for trainees to take part in the game, it is very rare that someone declines to play and be seen to refuse to "take the risk". Using this type of "fun" demonstration that involves and engages everyone present, is a very effective way of getting across your message.

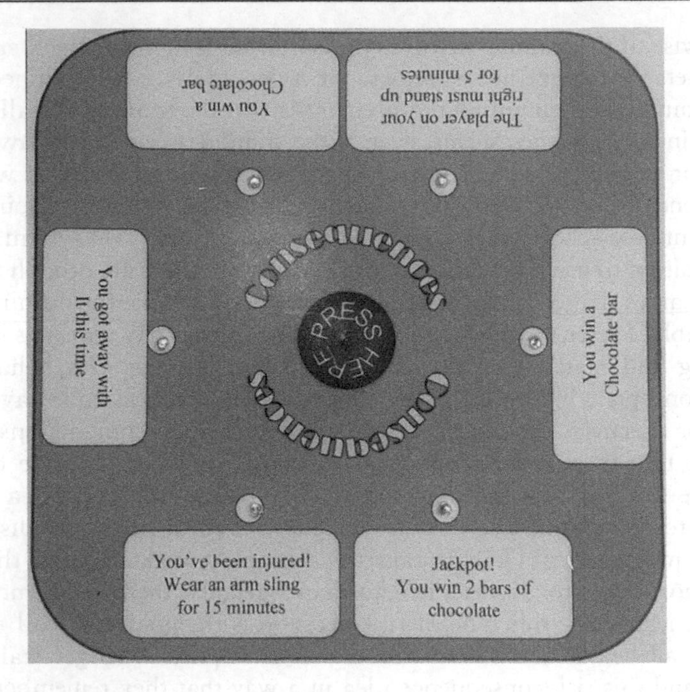

Figure 7.21 The consequences game.

You may not be able to find a proprietary game that you can modify to make this same point, but the same principle can be achieved by just getting trainees to throw a dice. I have also produced a version of the consequences game in playing card format. This is more controlled for you as the tutor, because it is not random – you can set up the cards to make the particular point. (For more details see the Resources Section F2.4 at the back of this book.)

7.14 EXHIBITS AND DISPLAYS AS AN AID TO TRAINING

Sometimes it can be helpful to use a display or small exhibition to supplement training. If this is done, it is important that it is relevant to the subject being taught and that the trainer should refer to it from time to time. It can be helpful to have some "freebies" on the display as people not only like to pick up items or information from displays, but the items taken back to their workplace will act as a memory cue and trigger not just about the item itself but also about the totality of the training. In the past I have designed mousemats (Fig 7.22) as a freebie and made them relevant to the topic being studied. These are a great memory cue particularly for people who regularly work at desk computers or control panels. In one case, I had trained an international group of managers about hazard identification and risk assessment. As a reminder of the six-step risk

Risk Assessment

Step 1 – Look for Hazards
Step 2 – Identify who might be harmed
Step 3 – Assess risks

CONSEQUENCE

LIKELIHOOD	Insignificant	Minor	Moderate	Major	Catastrophic
Almost Certain	Significant	Significant	High	High	High
Likely	Medium	Significant	Significant	High	High
Moderate	Low	Medium	Significant	High	High
Unlikely	Low	Low	Medium	Significant	High
Rare	Low	Low	Medium	Significant	Significant

Step 4 – Decide what extra controls are necessary
Step 5 – Record & communicate your findings
Step 6 – Review the assessment

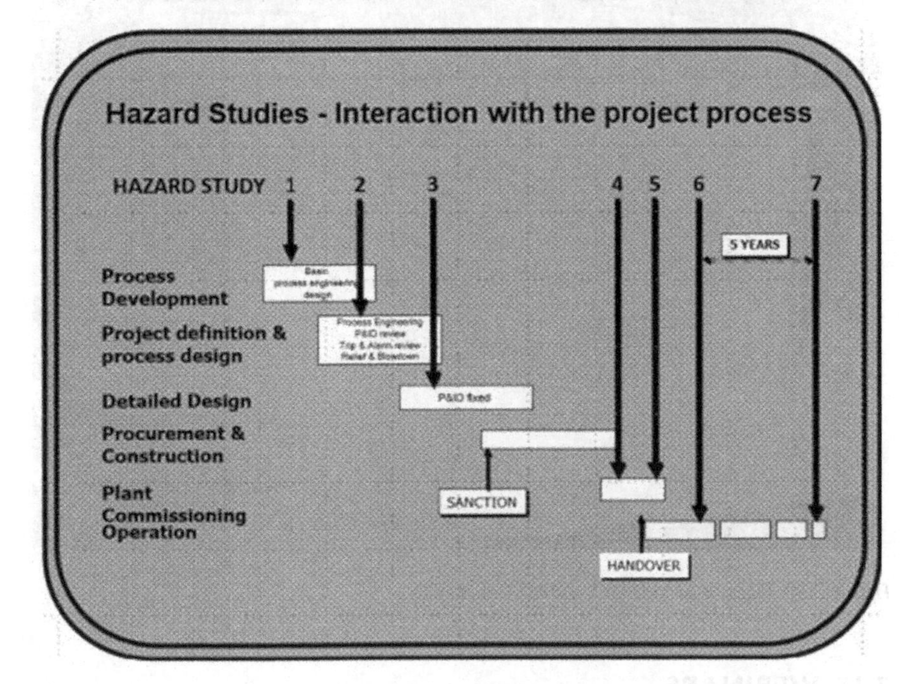

Figure 7.22 Examples of bespoke mousemats as a training aid.

management process, I gave them each a mousemat with the six steps of risk management and their organisation's risk matrix. Seven years later, I was carrying out a Safety Health and Environmental Management Audit at a factory in Germany and although I had forgotten about this particular individual at the factory, he had remembered me and proudly pointed on his desk to the risk management matrix mousemat that I had given him seven years earlier. It had been reminding him of the process that he had learned every day for the past seven years. He is unlikely to forget that.

Pre-prepared exhibition stands might also play a role in training. The one shown below is used to complement training in emergency and crisis management. That particular training is intended to be at an awareness level for managers in general, but some of the managers who attend have defined roles in their organisation's emergency plan and they are often wanting to know more detail. When this situation arises, I suggest to them that we explore the matter in more depth after the end of the training session and then we can use the display and its more detailed information to explain the matter further (Fig 7.23).

Figure 7.23 Picture of SHEEMS exhibition display.

7.15 WEBINARS

The Covid pandemic of 2020/21 accelerated the use of virtual meetings and remote training techniques. At the time of writing there are two principal systems in use for mass market video conferencing. These are Zoom from Google and Teams from Microsoft. Trainees in academic institutions may also use "Blackboard Learn" from Blackboard International B.V. and new software systems are being developed all the time. These systems have revolutionised how we get together and will have a permanent effect on our previous desire to always meet face-to-face. The use of video conferencing for training has some very obvious benefits. They are particularly beneficial when distance training people in different physical locations such as different countries or different parts of the same organisation. However, think about your own use of webinars. What is the ratio of webinars booked, to the number of webinars attended to the bitter end? You may well find that you have booked a lot more than those that you actually completed and if that is the case for you, it will be a similar ratio with your trainees.

In particular webinars are good for:

- Ease of setting up
- Reduction in travel time, risk and cost
- The ability to use a skilled tutor to train outside their normal workplace
- Getting a uniform message to multiple locations, avoiding the dilution effect

There are also some disadvantages:

- Reduced tutor-to-trainee interaction
- Questions are rare
- People tend to speak over each other
- Currently it is difficult to do practical work
- They tend to be done in a "lecture style" with limited opportunities for questions
- Can be problematic if people are unsure of the computer technology or the technology fails.
- It is difficult to be sure that participants remain focussed (or are even there!) when video is muted
- There is little chance for fun and "banter"

There are some differences in functionality of the "Teams" and "Zoom" systems. Zoom allows for a gallery display of all participants in the meeting which does allow for the tutor to get some visual feedback and to be able to see some body language and expression. However, both Zoom and Teams lose the participants images down to thumbnail size once the presentation is displayed. In the case of Teams, it is difficult to get a gallery of more than four or five participants after which time they are reduced to initials. This means that it is difficult for the tutor to know if the participant is actually there, particularly if their video and audio feeds are muted. Teams has a good system of queuing questions from participants and also has a facility for arranging virtual break-out rooms, but these are only of real use if the syndicate exercise is a discussion or brain-storming one and does not require any physical "props".

One way of increasing delegate involvement in webinars is to use the Mentimeter system from Menti.com This system allows the audience to use their smartphones to connect to the presentation, where they can answer pre-prepared questions to create fun and an interactive experience. The system aggregates the responses from multiple users almost instantaneously. It is slightly gimmicky, but can be useful for running quizzes or checking the delegates state of knowledge in real time. The first time you see it in use it has significant impact, but thereafter you know what is coming! The system is mainly beneficial to larger organisations who can afford to purchase membership.

The danger with webinars is that because they are easy to arrange, that in future they will become the default system for health and safety training. They can be very useful in some situations, but like any tool, they have their limitations. The key message when planning a webinar is to check:

1. Is this the best method of training in this set of circumstances?
2. Are there any participants who are uncommitted to the subject (if so they may need face-to-face training)?
3. The time for an effective webinar should not exceed 2 hours.

The 10 essentials of running a webinar presentation are:

a. Do not be too ambitious with the quantity of the content.
b. Make sure that you can be heard – buy a professional quality microphone.
c. Ensure that you have a fast and reliable internet connection.
d. Speak slowly and clearly.
e. Capture their attention with something memorable.
f. Observe the slide content etiquette described in the next chapter (i.e. not too much text/slide).
g. Insert regular visual highlights (funny picture, photograph video clip) at the stages where the attention might start to drift. I call these "attention grabbers".
h. Record the webinar for future use. But remember not to make it date or time dependent (i.e. do not put the date on the title page if you intend to re-use it).
i. Take time to respond to questions (typically there will not be many!).
j. Create a transcript of the webinar and make it available to trainees. This can be done by extracting the audio file from the recording and having that transposed into text.

Training Techniques and Styles Checklist

1. Trainees need to be engaged in the training. To achieve this, the training needs to be relevant and fun.
2. No single technique is the solution. The needs of the trainee will depend on:
   - The knowledge base (How much do they already know?)
   - Their demographic:
     - Younger people are more comfortable with computer-based training
     - Older people prefer more traditional styles
   - There understanding of the processes and terminology used at the facility
3. The key point is that trainees need to be involved and not just talked at. This will require that they learn by some practical experience.
4. Experiences and techniques might involve using one or more of the following:
   - What type of training is appropriate?
   - Videos as a training aid
   - Models
   - Posters or chart boards
   - Sticky note displays
   - Real-life experience
   - Simulations
   - Role plays
   - Photographs
     - Photographs in work instructions and training manuals
     - Photographs as posters
     - Photographs in safety training presentations
     - Photographs used in training exercises
     - Photo hazard spotting exercises
     - Photographic hazards spotting (Camera Hunt)
   - Jigsaws
   - Using Flip Charts
   - Drawings and wallcharts
   - Props and Gizmos
   - Exhibits and displays as an aid to training
   - Webinars

Chapter 8

Preparing computer displayed visual images – What you see is what you get!

8.1 POWERPOINT PRESENTATIONS

Most people nowadays are familiar with computer-based PowerPoint presentations. Many will also have experienced the feeling of being subjected to the torture of "Death by PowerPoint". PowerPoint is an immensely powerful tool, but as with any tool, it can be easily misused if applied by someone who has not been trained in its use and misuse.

The most common problems are:

- Putting far too much text or detail on each slide.
- Using too many slides.
- Overuse or misunderstanding the use of animation.
- Loss of control of the timing.
- Not knowing how to transfer between media.

A very common mistake is to try and put too much text on each slide. The problem arises that instead of listening to the tutor, the trainee is fully occupied reading the slide. Almost certainly the trainees reading rate and the tutors speaking rate will be different, and so the tutor will not necessarily be speaking at that moment about the text that the trainee is reading. This usually means that the trainee concentrates more on his or her reading, rather than what the tutor is saying. This often leads to misunderstanding or boredom. The solution to this problem is brevity of text and detail on the slide. The following extracts from training visual slides (Fig 8.1) show that with a little thought the impact of the information on the slide can be condensed without detracting from the meaning.

The slide on the left was provided to train a group in what operator safety checks to look for before using portable electrocal tools and their consequential hazards. To the observer it is a bit like looking at a page out of a telephone directory. Some people just will not bother reading it and others will be reading points 3 and 4 while the presenter is still amplifying point 1.

DOI: 10.1201/9781003342779-9

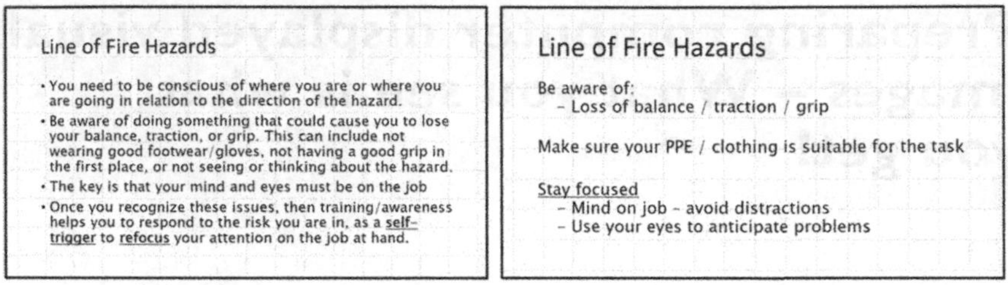

Line of Fire Hazards

- You need to be conscious of where you are or where you are going in relation to the direction of the hazard.
- Be aware of doing something that could cause you to lose your balance, traction, or grip. This can include not wearing good footwear/gloves, not having a good grip in the first place, or not seeing or thinking about the hazard.
- The key is that your mind and eyes must be on the job
- Once you recognize these issues, then training/awareness helps you to respond to the risk you are in, as a self-trigger to refocus your attention on the job at hand.

Line of Fire Hazards

Be aware of:
- Loss of balance / traction / grip

Make sure your PPE / clothing is suitable for the task

Stay focused
- Mind on job - avoid distractions
- Use your eyes to anticipate problems

Figure 8.1 Simplifying the content of presentational slides.

The right-hand slide shows the same basic information in a more concise way. The number of words has been reduced to a half but the impact of the information on the trainee is actually greater. This has not only been achieved by the reduction in words, but by the use of bullet points, indenting of text, underlining and selective colouring. Even now the information is still not easy to take in. It may be worth spreading the information over two slides. The impact of the slide could be further enhanced by the addition of photographs to show what a Portable Appliance sticker looks like, or to help people understand the typical cable faults that the user could expect to find. Remember that images like photographs stick in the memory much more effectively than text or the spoken word. Pictures and photographs have impact on the audience and tend to be remembered.

8.2 TOO MANY SLIDES

One of the dangers for inexperienced presenters is trying to put too much information into the presentation. This results in having too many slides to present. You should always practise a presentation prior to the first time that you use it. It is always best to do this with a trusted colleague. Many presenters forget to allow time for questions. Remember that if you get no questions, then either they know it all already or you are not having the desired impact and interaction. If you get no questions then things are probably not going well. To some extent, the more questions that you are asked, the more interested and engaged your audience will be. So, when testing your presentation with your colleague, encourage them to ask the questions that they would expect the average trainee to ask and ask them to give you feedback on whether your presentation is meeting the requirements of the training. This trial run will give you a good idea of whether your timings are realistic or not and you can also rehearse some of the answers that you would give to expected questions.

Every presenter's training delivery is different, but for me, as a very rough guide, I work on taking an average of about 3 minutes per slide. If you think that you can deliver a complete presentation at the rate of 1 minute per slide, then I would suggest that you are being over ambitious or alternatively you are not allowing time for the trainees to assimilate the information that you are delivering. Remember also that the timing of your presentation must also take account of the time for exercises, exercise feedback and natural breaks.

8.3 IMPORTANT TIP

The first time that you use some newly developed training then always have some "sacrificial content" in the training. This is material that if time starts to seriously overrun, that you can bypass or omit. Remember that trainees will forgive the tutor for many things, but the one thing that they will not tolerate is finishing late. Even if you think that you can get through the overrunning material very quickly, everyone else will have switched off and will have rising levels of irritation as they need to be away and get their bus or lift home.

Examples of sacrificial material:

- Video clips
- Exercises
- Games
- Plenary discussions

It is also possible to bypass parts of your presentation that you may consider to be of less importance and therefore in a time-limited situation can be "sacrificed". This is best done by putting in a discrete hyperlink to a future slide. It is best to have the hyperlink launch point disguised – so don't put in an arrow or button at the bottom of the slide, but disguise the link on something like a logo or title so that the audience do not recognise what you are doing!

Sacrificial material can also be removed or bypassed while the participants are otherwise engaged in syndicate exercises or the like, where their attention is away from the tutor and display screen. Again, in order to make this alteration discrete, ensure that if you are skipping slides, always mute the image on the display screen. It looks really unprofessional if the participants can watch you put up the slide sorter which not only shows that you are skipping some of their training, but also shows them what else is to come.

Be aware that there is a danger with this sacrificial approach. If you have provided the participants with sequential handouts that refer to the sacrificed material, they will know what you have done and may ask awkward questions!

When presenting a particular training session for the first time, it can be just as difficult to anticipate an under-utilisation of time as it is to anticipate over-runs. I was asked once to do a presentation to a large group of client employees. I had been told that the training event was to be 2 hours long and that my session was limited to one hour. When I arrived for my input, I was told that the other speaker had pulled out and that to avoid embarrassment of the organisers, I needed to fill the full 2 hours. Luckily, I always carry additional material, but instantaneously filling an extra hour is quite a challenge. I was able to use some of my pre-prepared group activities to fill the space and the organiser's embarrassment was avoided. Although finishing hours early looks as though you don't know what you are doing, finishing a few minutes early always gets appreciated. In most cases of significant under-run, I would recommend taking an extra or early coffee break, or for the tutor to give personal examples of experiences that relate to the subject matter, as these are always appreciated.

8.4 ANIMATION

The effectiveness of slides can be further enhanced by the judicious use of animation. This allows the presenter to bring up the points on the slide as he or she is talking about them, hence ensuring that the trainee's attention is focused on that section of the slide that the tutor is talking about.

Animation can easily be abused. Often the default setting on computers is for the text to "Fly in". On this setting the entire text literally flies in from the edge of the screen. It is impossible to read any of the text while it is "flying" and is a real problem to viewers with a propensity to migraine. This animation should never be used. Likewise, some of the spiralling effects are also impossible to read until the animation has stopped. My recommendation is to use one of the following animations:

Appear

Blinds horizontal

Wipe

If using "Wipe", then always ensure that if wiping text that the text wipes in the direction that the viewer will be reading the words. This is normally to wipe from left to right. The wipe animation is also particularly valuable when displaying block or flow diagrams. Flow diagrams are often a very good way of simplifying quite complex sequences or processes. However, the completed flow diagram can look quite complex on a display screen. In this situation, the presenter should animate the diagram so that it appears one block at a time. When exposing blocks, the wipe animation should be in the direction that the arrows point. Remember to "group" the arrows and flow diagram box so that they all appear at the same time (see Fig 8.2).

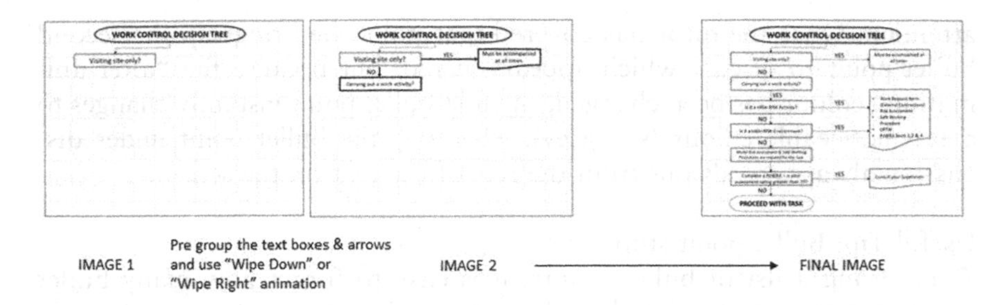

Figure 8.2 Revealing flowcharts progressively to maintain focus on what is being discussed.

Be careful about animating tables where the cells in the table might auto-size. This can occur where a table is populated with text and during animation may result in the shape of the table changing while the viewer is looking at it. Often part of the table may disappear off the side or bottom of the screen. The solution to this problem is to populate the table on a slide as you wish it to be seen at the end. If you wish to introduce data in five animations then duplicate five copies of the slide and then change the text colour on each slide to match the table background colour to effectively eliminate the data that you don't want to appear at each animation. The animation effect is then achieved by simply advancing to the next slide (Fig 8.3).

Hazard	Acute effects (short term)	Chronic effects (long term)
Asbestos	X	✓
Carbon monoxide		
Nitrogen		
Lubricating Oils		
Noise		
Manual Handling		
Legionella		

SLIDE 1 Most of cell text is coloured white

Hazard	Acute effects (short term)	Chronic effects (long term)
Asbestos	X	Pleural plaques, asbestosis, lung cancer, mesothelioma
Carbon monoxide	✓	✓
Nitrogen		
Lubricating Oils		
Noise		
Manual Handling		
Legionella		

SLIDE 2 Same table but some of text is revealed

Hazard	Acute effects (short term)	Chronic effects (long term)
Asbestos	X	Pleural plaques, asbestosis, lung cancer, mesothelioma
Carbon monoxide	Dizziness, weakness, nausea, collapse ("chemical asphyxia"), death may follow.").	✓
Nitrogen		
Lubricating Oils		
Noise		
Manual Handling		
Legionella		

SLIDE 3 etc., but more text is revealed and symbols overwritten

Figure 8.3 Example of table animation.

This technique of animation is not limited to tables.

Another useful facility in animation is the ability to change text colour. This is useful if the tutor wants to hold the participant's attention to a particular bullet point that they are discussing. If the tutor has a line of six bullet points that he has animated to come up one at a time, then by using the "Effect Options" tab in PowerPoint and then going to "Text Animations" and then "After Animation" we can choose to change the text colour after the next mouse click. What this means is that as the first bullet point is brought up, it might appear in red text to catch the viewer's attention. Once the tutor has covered that point, he brings up the second bullet point to discuss which appears in red, but because the "after animation" colour has been changed, the first bullet point instantly changes to a less noticeable colour (say grey). This way the bullet point under discussion always stands out from the rest of the text on the screen.

Useful Tip: Bullet point stop
When using a list of bullet points, it is easy to forget how many bullet points are on a particular slide. It looks as though you are not properly prepared if you are already on the last bullet point and you accidentally advance to the next slide, resulting in the need for you to apologise and go backwards. To avoid this happening only place a full stop after the final bullet point. This way the presenter knows that the bullet with the full stop is the last one on that slide and that the next click will advance the slide.

Bullet point lists on slides should therefore look like:

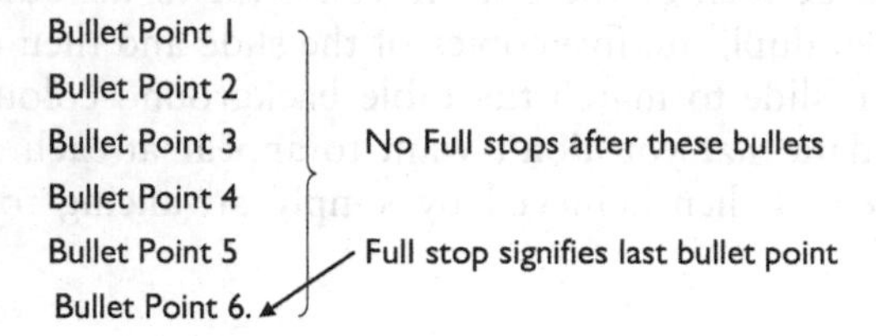
Bullet Point 1
Bullet Point 2
Bullet Point 3
Bullet Point 4
Bullet Point 5
Bullet Point 6.
No Full stops after these bullets
Full stop signifies last bullet point

8.5 MEDIA TRANSFERS

Using a multi-media approach to training provides variety for the participants and increases the impact of the training and therefore its memorability. The problem is that Murphy's Law always applies – if anything can go wrong, it will! Failure of some fancy media input, be it music, video, CD or an internet link can sometimes be a show stopper, but always looks as though the presenter is not competent. The key message is to test your multi-media inputs both during the training development, but also immediately before the presentation. The problem is although computer manufacturers claim that their computer screen displays are WYSiWYG (another acronym meaning "What You See Is What You Get") and that anything that appears on the computer screen will be duplicated on the display screen. Do not believe them! Experience shows that sometimes computers and displays are not compatible. This is especially the case with LCD projectors. Even if you have checked before, always re-check that the image projection works. Even if it does, you may also need to change the aspect ratio of the screen (is your computer set for

4:3 picture size or widescreen as this can distort your photos or video?) You should also check that the sound is connected and working and set at the right volume level. Ensure that the ambient lighting is not too bright if you are showing video clips. I normally recommend that videos are embedded in the presentation so that they start automatically on a mouse click. If you want to automatically return to the next presentation slide after the video has played set the slide advance on the toolbar to automatic after "X" seconds, where "X" is the play length of the video. However, if that is not possible, make sure that you not only know how to start the video, but that you also know how to return to the next slide in your presentation, as using the escape (Esc) key will return you to the slide sorter screen and not to the next slide in the presentation. I usually recommend professional presenters to have two computers available and ready to use, each with its own copy of the training material and presentation on its hard drive just in case one computer has problems. In some circumstances, particularly when using portable LCD projectors, it is also worth having a spare projector available. Another problem can be caused by the "computer to display" connector cables. Modern systems use WiFi or HDMI cables which are a digital link and now most laptop computers only have HDMI or USC ports. This is a particular problem for older training facilities, which I find tend to be in factories. Here many of the LCD projectors still use VGA (multi-pin) connectors. These operate using an analogue signal. To the best of my knowledge there is no way to adapt a VGA cable to connect to an HMDI port. So, check in advance what type of computer connection port is required to link to the display system that is to be used, because that may determine which computer you need to take with you. The backup to this dilemma is to always have an additional copy of your presentation on a memory stick, so that if all else fails you can insert that into the local computer system which is invariably in the training room and will almost always have a spare USB port to take your memory stick.

Using internet links during a presentation is also fraught with potential problems. Links go down or freeze and re-booting can take time, during which the trainees have nothing to do while you are sweating away trying to get the desired website back and trying to still look in control. The other problem is that internet usually relies on a WiFi link and if it is switched on, then personal WiFi access is also likely to be available for the participants, creating a diversion and lack of attention to the presentation. I would always recommend using internet access only in syndicate exercises or in one-to-one training.

Preparing Computer Displayed Visual Images, Checklist

1. The main common problems are:
   - Putting far too much text or detail on each slide
   - Using too many slides
   - Overuse or misunderstand the use of animation
   - Loss of control of the timing
   - Not knowing how to transfer between media

2. Speak to the slide – do not just read it out.
3. Use visual images, such as graphs, block diagrams and photographs, to enhance your presentation.
4. I find that my slides take an average of 3 minutes/slide to deliver. Your time may be slightly different (but not greatly so!).
5. Have some "sacrificial content" in your presentation that you can exclude if time becomes a problem.
6. Be careful about the use of animation:
   - Use animation to ensure that trainees focus on what you are talking about and are not reading ahead
   - Use only the "Appear", "Blinds horizontal" and "Wipe" animations.
   - Be careful with populating tables on the screen, as the cells can auto size leading to viewing problems.

7. Make sure that any embedded video clips and their sound systems are tested before every presentation as different training set-ups and projectors behave differently.
8. Always practise your presentation before the first training event, allowing for time for questions.

Chapter 9

Handouts and supporting material – Providing reference material

If the training is to be effective, it is important that it has a long-term impact. We know from Chapter 2 that being human, people quite easily forget things. The best way to help long-term memory is to provide trainees with some sort of supporting material that is usually written down. These are generally known as "hand-outs".

When preparing written information to support a training activity, it is important to consider the target audience. A handout containing text descriptions may be appropriate for a group of managers, but if you are training technicians in how to choose, fit and wear a respirator, then pictures, photographs, action lists and wall posters may be more easily understood and more memorable.

For conventional classroom-style training sessions, there are two distinctly different types of hand-out. The first is a document that contains information necessary to allow the trainee or group of trainees to take part in some sort of practical exercise. It is in fact a briefing note that might explain what they are required to do or provide them with background details for the purposes of carrying out a syndicate task or the like.

This type of hand-out needs to be a balance of being concise but at the same time conveying all the information necessary to allow the trainees to do the task. Often it can be very useful to use photographs, sketches and tables in this sort of handout. Remember that the purpose of the handout is not to test their reading ability, but to provide information. An example is shown in Fig 9.1.

The second form of handouts is those which are provided for the trainee either before or at the beginning of the session. The purpose of these handouts is to provide a memory jogger in future to remind the trainee of the salient points of the training. They can also be useful as somewhere for the trainee to append explanatory handwritten notes. Providing good quality handout notes to support a training event can be very time-consuming, but are also essential in providing long-term reference material for the training to use throughout his or her career.

DOI: 10.1201/9781003342779-10

Manual Handling Exercise

Operation – Secretary reaching for box file

An office worker is lifting the box file from a shelf

- The office worker is stretching from a sitting position
- The action requires twisting through 60 degrees and the chair is not a swivel chair
- The task is done 20 times each day (10 times retrieving file & 10times replacing file)

TASK

1. Using the wall chart assess whether this operation is likely to lead to a musculoskeletal injury
2. Produce your responses to each task on a flip chart and be prepared to report back to the other delegates.

Figure 9.1 Example of a manual handling exercise brief.

The easiest type of handouts is based on providing miniature versions of the PowerPoint presentation slides. These can be produced directly from the PowerPoint file either by using the commands:

- Go to "Publish" then select "Create Handouts in Microsoft Office Word"

Alternatively to print directly from the PowerPoint file:

- Go to "Print" (Ctrl + P)
- Under "settings" box change from "Full page slides" to "Handouts option 3" which will print out the slides at 3 to a page with notes space.

Either option should give you handouts in the format shown in Fig 9.2.

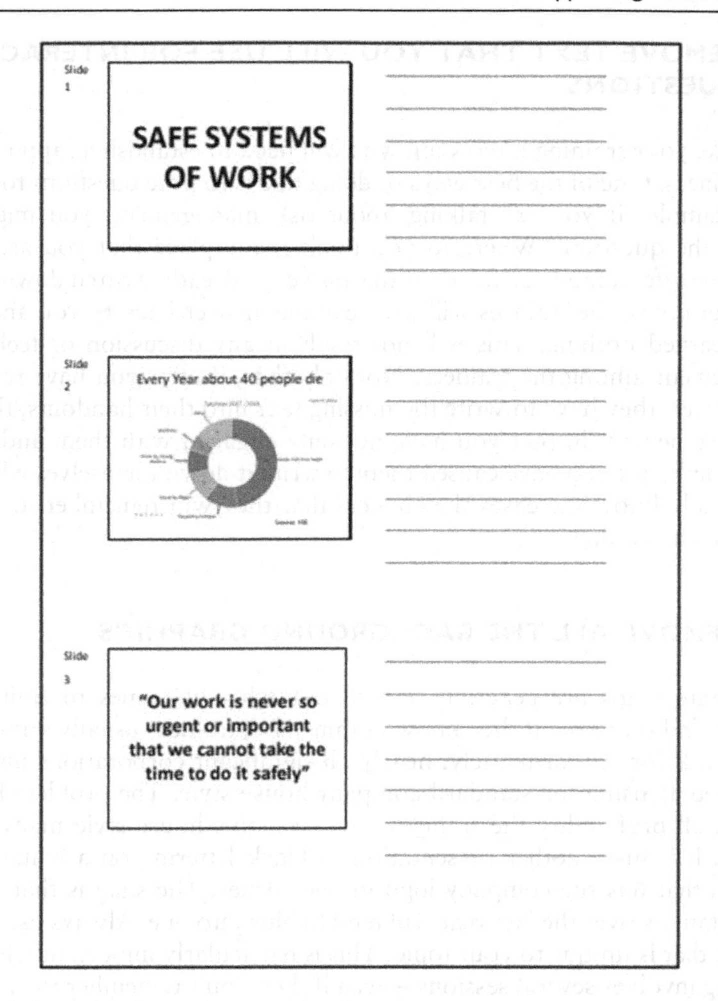

Figure 9.2 Example of typical trainees handouts.

If the handouts form one section in a multiple section training event, then always ensure that each new set of handouts starts at the top of a page. If the last handout in any session does not complete a full page of three miniature slides, then add either one or two blank slides into your handouts presentation so that the pages can be automatically printed in multiple batches without you needing to be present at the end of each sessions print run.

It is essential to keep a separate electronic file containing your handouts for each training session because it will differ from the file that you use for your presentation. The easiest way to do this is to save a copy of your training presentation – to identify it as the handouts master file. I tend to add the suffix "HOut" to the original file name. Before printing the handouts, you will need to edit them.

9.1 REMOVE TEXT THAT YOU WILL USE FOR INTERACTIVE QUESTIONS

To make your training interesting, you will need to establish a rapport with the trainees. One of the best ways of doing this is to pose questions to them. For example, if you are talking about risk management, you might be asking the question "Where to you think is the place that you are most likely to suffer a fatal accident". If the answer is already written down in the handout notes, the trainees will just read the answer back to you and they have learned nothing. This will not result in any discussion or feeling of involvement among the trainees. Not only that, if after you have revealed the answer, they have to write the missing text into their handouts, there is a double benefit, in that you have not only engaged with them and made them think, but you have caused them to write it down themselves which as we already know increases the chances that they will remember it. It also keeps them awake!

9.2 REMOVE ALL THE BACKGROUND GRAPHICS

Corporate logos are generally a bad idea when it comes to training. I usually find that even if they know nothing else, trainees usually know who they work for. Unfortunately, nearly all significant corporations insist on their people using the standard company house style. The problem here is that in all probability the trainees have seen this house style many times before. It is just another presentation of black lettering on a white background that has the company logo in one corner. The snag is that all the presentations over the last year will tend to blur into one. Always use a slide format that is unique to your topic. This is particularly important when the training involves several sessions – even if they don't remember the detail it can help to remember that was the session with the yellow background!

The images of your slides in the handouts will be quite small, and the text will be even smaller. It is essential that the text is readable and this is one of the reasons why you should not put too much text onto each PowerPoint slide. To create the maximum amount of space on the handout slide, any unnecessary graphics should be removed before saving the handouts file. This includes the removal of such things as clip art, background graphics and logos.

Clip art can be deleted slide by slide, but background graphics and logo can be removed for the whole file in one go. To remove these, go to the PowerPoint toolbar on the top of your screen:

- Go to the "View" tab
- Click the "Slide Master"
- Click the "Hide Background Graphics" box
- Review the handouts to ensure that the text is large enough to read

If you are planning to duplicate multiple copies of the handouts, as is usually the case, it is useful to do it all in black and white against a white background. If you want to reproduce photographs in the handouts and you have set to print in black and white, remember to click the "Greyscale" box.

9.3 MAKING CHANGES

It is quite usual to need to amend the main PowerPoint presentation from time to time to respond to feedback from previous trainees. Do not forget to go back and also make the same changes to the handout file, otherwise your handouts will not match the current version of the presentation.

9.4 HANDOUTS FOR INTERNATIONAL AUDIENCES

In this age of global organisations, it is becoming more common that training courses are attended by trainees who do not all have English as their first language. The tutor must recognise that even trainees who have a good standard of English as their second language, find it very challenging and tiring to concentrate on being taught in a language that is not their own first language. If their concentration lapses for a moment, they can lose the thread of what is being said and they can easily get lost. One way of helping them get back to understanding what is going on is to provide your handouts as a translation that they will understand. This is not as expensive as it sounds and is a great help to foreign trainees. Just a word of caution – even professional translators make mistakes. In one case that I used, the words "Chemical Plant Process Safety" were translated as "Chemical Daffodil Process Safety", which didn't exactly help understanding! When you have had a translation done, always get someone who speaks that language to read it for obvious errors before you publish it. Don't be tempted to use computer-generated translations – they never work!

9.5 PROFESSIONALLY PRINTED HANDOUTS

I am sure that we all have files in our offices that contain handouts from training courses that we attended years ago, but have never opened since. Often these old training files are quite extensive but not easy to revisit or search when you want to refresh your memory on a particular aspect of the training. I find that handwritten and computer-printed notes tend to get lost or destroyed over time, whereas books and professionally printed notes have a much longer life expectancy. Professionally produced handouts can take time and expense to produce and so are not appropriate for all circumstances, but for consultants doing repeat training events or for large-scale training needs, the use of professionally printed handouts may be appropriate and carry an air of authority that computer printed handouts may lack. The important thing is to keep this type of handout short and punchy with easily accessible indexing or contents lists, so that they become a really useful reference handbook.

The drawback of professionally printed handouts/handbooks is that they are very difficult and expensive to change – so it is essential to get the content right before committing to the print. This means that always run an initial pilot training event and get feedback before finalising the print proof. If considering professional publishing for extensive and detailed handouts, it is worth considering ring style binding so that individual pages can be updated and replaced without the need for wholesale re-printing.

Examples of several different types of bespoke professionally printed safety training handouts are shown in Fig 9.3.

Computer printed handouts

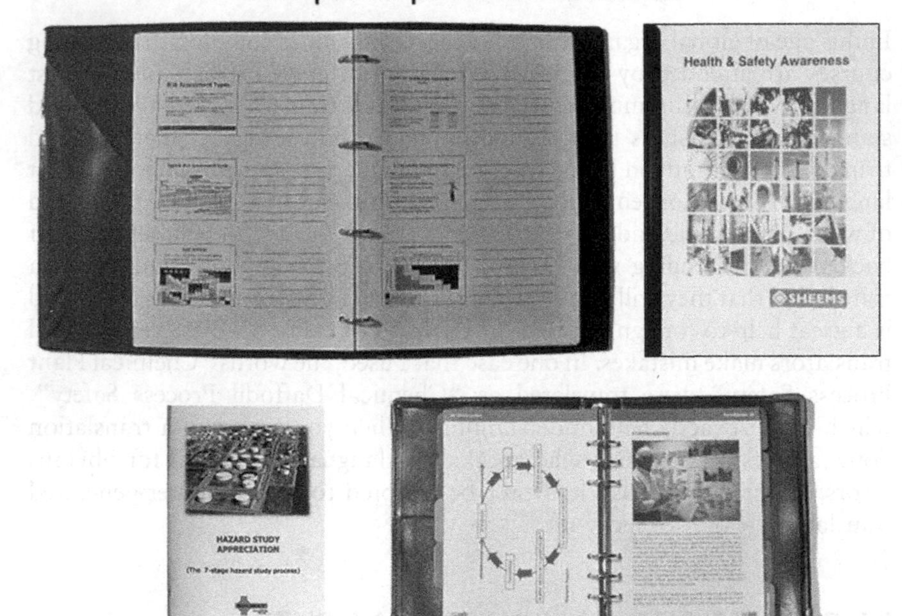

Professionally printed handouts

Figure 9.3 Examples of printed handouts.

9.6 USING HANDOUTS DURING A PRESENTATION

The tutor should always have a copy of the handouts available for his or her personal use. It is important to refer to the handouts periodically during the training session to make sure that the trainees are following the session and everyone is on the same page! Always check the copies of the handouts before you give them to the trainees. I was attending a week-long training course in Belgium some years ago and we were all presented with a very large course manual of handouts for the event. Once we got started, the manual became impossible to follow as there were about 150 pages in a totally random order. It appears that an inexperienced office intern had been asked to print the duplicate copies of the course manuals. Somehow this person had got the pages all mixed up and not only had not realised that, but in any case would have not had the knowledge necessary to correct the mishap. Always check manuals and handouts before the training starts. I find it a good idea to insert page numbers to make this checking process easy.

The tutor's own copy of the handouts can also be used to mark up any changes/learning that may be relevant to future applications of this training presentation. Trainees will always appreciate it if you recognise their good ideas or knowledge and see you making a note to share it with other future training events.

9.7 BOOKMARK HANDOUTS

In some cases, it may be appropriate to produce a cue or prompt card that is small enough to fit into the trainee's pocket or notebook, so that they have a reminder with them in the workplace of their training key points. The most convenient of these is having the prompt card in the form of a bookmark. Such a card is very limited in size and so can only hold bullet points but this can be very valuable when carrying out behavioural safety discussions, accident investigations or audit checks. The cards have to be fairly robust and should be printed on good quality card, such as photo paper and then laminated before being cut to size. Fig 9.4 shows an example of a bookmark handout used as a reminder for audit checks.

AUDIT BOOKMARK

Preparation:
- Read the procedure
- Who is responsible?
- Arrange meeting

Checklist:
- Who
- What
- When
- How

Discussions:
- "Show me how you....!"
- "Why that way?"
- "What do you do?"
- "How do you ensure that..."
- Listen

Observations:
- What do you see?
- What is missing
- Look:
  - Above
  - Beneath
  - Beyond
  - Behind

Documents:
- Complete?
- Up-to-date?
- Changes
- R-SAC:
  - Reason
  - Select
  - Assimilate
  - Challenge

Reporting:
- Significant issues
- Based on evidence

Figure 9.4 Audit checks bookmark.

9.8 PONDER SHEETS

Training is not a one-off experience that is limited to a particular training course or location. The training must be durable. The whole idea is that the trainee not only learns and remembers the training, but that he or she puts that training into practice when they return to their workplace.

Training often involves the application of some general principles and the application of those may depend on where the trainee works and their job role. Although both office staff and factory workers might undergo similar risk training, the way in which they will apply that when they return to their daily jobs will differ, because of the very different hazards that each is exposed to. It is often not possible to cover every detailed application of the training within the training event itself and so the trainees need to identify for themselves what actions they are going to do differently when they return to their workplace following the training. One way of getting the trainees to prolong their learning after the training course has ended is through the use of "Ponder Sheets". These are designed to make the individual reflect on the training and personalise it for themselves. They do this by indicating:

1. What was the most important message for me that came out of the training session?
2. What new actions do I need to take when I return to work as a result of what I have learned?
3. Do I need to find out more or get further training on this subject; and if so,
4. How do I go about getting more information/training?

Training Course

PONDER SHEET

Training Session Topic
What is the most important message for me personally to come out of this session?
Are there any actions that I should consider taking on my return to work?
Do I need to know more about this subject? YES / NO If "yes" how should I go about that?

The value of the use of Ponder Sheets is that it recognises that people come to a training session with different levels of knowledge and experience on the subject being covered. One of the problems for the tutor is knowing at what level to pitch the training. If it is too simple, the experienced people will be disengaged and if it is too advanced, the newcomers will fail to grasp the basics. The Ponder Sheet allows for trainees to recognise for themselves what is most important and by doing so, they are more likely to remember the training and apply it.

Ponder Sheets are usually incorporated with the training handouts manual and are particularly valuable in general safety awareness training where there are a number of different aspects of health and safety are being learned over a more prolonged period of time.

Training Handouts Checklist

1. The best way to help long-term memory is to provide trainees with some sort of supporting material that is usually written down. (Handouts)
2. There are two main types of handout:
   - Information which supplements the tutor's explanations. These are usually provided prior to or at the start of the training event. They are often in the form of a loose leaf binder.
   - Information that enables the trainee to undertake some sort of practical exercise (i.e. a briefing note or instruction). These are usually handed out at the time they are needed.
3. When preparing handouts from PowerPoint presentations remember to:
   - Remove text that you will use for interactive questions
   - Remove all the background graphics and clip art
   - Ensure that the final handout is legible
4. What is the best format for the handouts?
   - Loose leaf folder (A4/Foolscap)
   - Professionally printed A3 booklet (good for long-term reference)
   - Professionally printed loose leaf folder (good for consultants delivering the same courses repeatedly with minor updates)
5. Consider including "Ponder Sheets" within the handouts in order to make trainees think about what is important about what they have just learned.

Chapter 10

The art of presentation delivery – Getting your message across

History in the form of that great Christian reformer Martin Luther told us that a good preacher should have these properties and virtues:

1. Teach systematically.
2. Have a ready wit.
3. Be eloquent.
4. Have a good voice.
5. Have a good memory.
6. Know when to make an end.
7. Be sure of his doctrine.
8. Venture and engage body and blood, wealth and honour, in the Word.
9. Suffer himself to be mocked and jeered of every one.

The health and safety trainer of today does not need to be a cleric, but the qualities that Martin Luther advocated are still relevant today, in that systematic teaching, wit, eloquence, being sure of his doctrine and most of all knowing when to end, are as relevant to speakers of today as they were in the 16th century.

10.1 GIVING A PERFORMANCE

What many health and safety presenters forget is that making training presentations is all about giving a performance. If the audience do not find what you are saying and doing engaging, then your message will be lost. The real challenge with health and safety training is that nearly everyone agrees that it is important, but after that it can be a very dry subject. The tutor needs to capture and retain the trainees' attention. The four aims of the tutor are designed to give the trainees an EDGE with their newfound knowledge and skills. In this context the EDGE acronym stands for the four main aims of the tutor and stand for:

DOI: 10.1201/9781003342779-11

Educate
Demonstrate
Guide
Excite

There are three phases in any presentation which are usually summarised as:

- Tell them what you are going to tell them (Explain why we are here)
- Tell them (The main learning)
- Tell them what you told them (Summarise the key messages)

10.2 THE DIFFERENCE BETWEEN DATA, INFORMATION AND KNOWLEDGE

Learning is about providing knowledge and understanding to trainees which relates to their particular needs. Often the starting point is an array of raw data as portrayed in Fig 10.1. Without the ability to understand the data it is of little value. Data needs to be interpreted and put into a meaningful context in order to for it become of real value. The tutor must take the raw information and present it to the trainees in a way that they can see the relevance of it to their situation and needs. For example, accident statistics are initially just a set of numbers. In order for them to have value and be acted upon, the raw numbers need to be analysed so that the

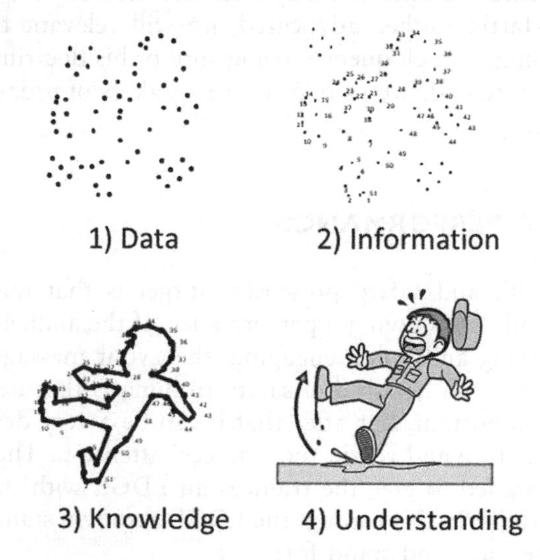

Figure 10.1 The difference between Data, Information and Knowledge.

management team or trainees can learn where, why and to whom accidents are happening. It is only then that it is possible to decide on some corrective action. This process of analysis changes the raw data into useful information. In the drawing no. 2 the raw data has been changed by adding numbers to some and leaving those that are irrelevant, unnumbered. This shows that analysis of the data has identified that some data points may be more significant than others. The data has now become potentially useful information.

In a training situation, the tutor will use information as evidence that there is a need for training. However, the learning objective is to help the trainees to "join up the dots" and see beyond the information, so that they learn the broader messages behind that. Once the trainees have been given the tools to deal with the information, they have acquired new knowledge which in the case of drawing no. 3 allows them to "join up the dots" and start to see the shape of what the data is telling them.

The final aim of the training is to get the trainees to not just join up the dots and apply their learning automatically, but to anticipate what the information is telling them about a particular aspect of health and safety in their workplace. The aim of a fully trained and competent worker is that they learn and fully understand why they have to behave safely. This means that they do not just respond to things that go wrong, but that they understand and anticipate problems (drawing no. 4) and are able to avoid them.

Remember that the problem of being overwhelmed by data is not just a numbers problem. The same issue arises when tutors overwhelm trainees by their use of words. The words that you use can often be meaningless jargon to some trainees and so we must always convert our words into information that the trainees can appreciate and understand. This is particularly important when dealing with individuals who may not have English as their first language. Some of the western health and safety regulatory agencies are now producing visual training packages to aid understanding in some of these multi-lingual workforce training events.

10.3 ADDRESS TRAINEES BY NAME

It is always important to be able to address trainees by name. If you are leading a training session that has people that you don't know, always have name cards or name badges available. The best form of name card is an A5-sized plain white file card folded in half longways to form a "toblerone" shaped card that will stand alone on a desk. Decide whether you want full names or just known Christian names. Ensure that names are written in dark marker pen and not pencil or biro, so that you can see them easily at a glance. If some of the trainees are unknown to the others, then it can be useful to have names on both sides of the toblerone. I usually leave it to the participants to put their names on the card. However, if you wish to put certain people together (i.e. a mix of experienced people with newcomers) then write the name cards in advance and place them where you want them before the trainees arrive. This can be particularly helpful if you anticipate putting two friends together might lead to problems or lack of attention!

If the attendees do not know each other or do not know you, then take the time to let everyone briefly introduce themselves. Equally if there are individuals attending who do not know you, make sure that you introduce yourself and provide a brief introduction as to why you have been asked to lead this training event.

10.4 DRESS CODE

The tutor needs to present him or herself as professional but approachable. I like to dress to suit the occasion. A tutor turning up to a maintenance workshop in his best Sunday suit, to train a long-in-the-tooth mechanic is not likely to be viewed by the mechanic as very credible when it comes to practical matters. In these circumstances, convey the impression that you are prepared to get your hands dirty and come wearing clean overalls. On the other hand, turning up in overalls to educate the company board members on some aspect of health and safety policy will also be frowned upon and a suit might be more appropriate in those circumstances. For most "off the job" training courses then smart casual dress is normally the safe option. Avoid wearing something that is so eye-catching that the trainees' attention is distracted from the primary subject of the session. If you are arranging an off-site training session, remember to state the dress code in the course joining instructions, particularly if it involves some practical on-site work that might require them to bring their own personal protective equipment. If the training involves working at heights on grating-type flooring, then I would also want the ladies to know in advance to wear trousers.

It is very easy with health and safety training to inadvertently send the wrong messages or undermine your own personal credibility. It is crucially important to ensure that as you arrive for the training session and during it, that you wear the appropriate personal protective equipment and follow the local safety rules. If you are training on a construction site, wear a hard hat and safety footwear as soon as you step out of the car. You will be astonished how much people notice, particularly if they think that they can embarrass you later on! You can hardly be asking them to wear light eye protection, if you have not been doing so yourself! So do not inadvertently forget to observe some crucial piece of safety regulation that might set the training and your own credibility off on the wrong foot.

Finally, before you start, make sure that all appropriate zippers and buttons are done up and shoelaces are tied to avoid having to deal with subdued sniggers.

10.5 MANNERISMS

Whether we like it or not, we all have mannerisms. Some of these can be particularly annoying or distracting to those trying to listen to what we are saying. Try and identify those in advance so that you can avoid them if at all possible. Particularly annoying mannerisms might be:

- Jangling keys or coins in your pocket
  - Solution – Make sure that your pockets are empty before you start.
- Saying the word "Err..." repeatedly
  - Solution – Make a recording of yourself speaking and see how often to say "Err". Repeat the recording over several times until you can speak without "Err...ing".
- Smoking
  - Solution – In most locations in the West, smoking inside is not permitted. Make sure that you allow smoke breaks for those people who need it during the training session.
- Mobile phone calls
  - Solution – Ask everyone (including yourself) to switch off/turn phones to silent.
- Physical distractions like "tics" or "shakes"
  - Solution – These are often a sign of nervousness. The more you do, the less nervous you will get. Practice the presentation beforehand. A good way of checking if you are susceptible to this is to video record some of your presentations and play them back to critique your own performance.
- Broad accents
  - Solution – Slow down your delivery. Warn people in advance and ask them to tell you if there is anything that they do not understand.

Finally, watch out for the booby traps around the room. The legs of flip chart stands and electrical power leads are notorious for being trip hazards. Wherever possible walk around the training room before the trainees arrive and identify any obvious hazards. In particular, it is essential in a health and safety training session to ensure that extension cables are either taped down or in a proper cable protector sheath. Such tape and cable protectors should be a standard part of the health and safety trainer's equipment (see the Stationary box – Chapter 13).

10.6 ENTHUSIASM

Nothing is so infectious as enthusiasm. The subject of health and safety can be rather dull, so it is your job to inspire people. I was running a health and safety training workshop for international managers. At the end of one of the sessions about "learning from accidents" that seemed to have gone quite well, one of the attendees came to see me afterwards. He had a big grin on his face and said to me "You enjoyed that, didn't you!" I thought for a moment and realised that he was right – I had enjoyed it. If you enjoy the training, the trainees are more likely to enjoy it and they are much more likely to remember it.

It is almost impossible to appear enthusiastic if you are sitting down or standing behind a lectern, like a vicar reciting his sermon or a politician on a podium. I always encourage presenters to move around and smile a lot. Go up to people and look them in the eye. If someone answers your question approach them with your further response, so that it is clear that you are talking to them personally and not just broadcasting a message. The number of trainees and the layout of the room is really important in allowing you to approach people personally. It is easy to approach and engage with people sitting in a horseshoe layout, but it is much more difficult in a theatre style or classroom layout.

If you are enthusiastic, then trainees will remember both the message and you. Remembering you is helpful the next time they attend one of your training sessions, because they will come along expecting to get something useful out of it.

To be enthusiastic requires energy. It is difficult for a tutor to sustain very high energy levels throughout a long training session and so it is equally important to vary the pace of a presentation. Having raised the pace, if you are wanting to create pathos while you recall a real-life incident and its consequences, then slow down the pace and lower your voice to match the seriousness of your message. Do not attempt to make any light-hearted comments when talking about fatal injuries or things that have gone badly wrong. Choose the timing of recalling these high-impact events so that to cause the maximum impact you leave people thinking about what happened as you come to a natural break in proceedings such as a coffee break or the end of the day. I like to leave people completely quiet for a few moments before I almost whisper "whilst you think about that, let's have a few minutes break".

10.7 ENGAGEMENT

Engagement is all about getting the trainees to listen and respond to what you are saying. Without engaging with the trainees, the message will be lost and the training will be largely ineffective. It is important that each trainee thinks that you are talking directly to him or her. To achieve this, then the numbers of trainees and the layout of the training facility are important. If there are too many present or some people can hide at the back, it will be very difficult to engage with them. It is important to approach people and look them in the eye to get their attention. Sometimes it will be enough to scan around the room (called the lighthouse technique) to convey the message that you are talking to everyone, and on others it will be more relevant to address a particular individual. If there has been no feedback for a little while, get their attention by addressing someone directly by saying something like "Do you understand that, Steve?". Even a simple "Yes" answer has ensured that Steve remains engaged.

Avoid continuously looking at one person for long periods of time – they might feel "picked on" or others might think that you are ignoring them. Remember that as you move to a next phase of the training, to give everyone the opportunity to ask questions.

It is quite easy to check whether trainees have engaged with you. If you get to the end of a particular session and no one other than the tutor has spoken, then you can be assured that they were not engaged. I prefer to have lots of banter going on, because that way I know that trainees are engaged and thinking.

One possible downside of getting people engaged, particularly in the horseshoe seating layout is that you can end up with your back to the others and they might not hear what you are saying or feel left out. Remember while answering an individual's question to scan (the lighthouse technique) the room to make sure that other people feel included in your response.

If the training is likely to continue for a day or more and has little practical content, using more than one tutor can help to change the voice tone and introduce variety and a different style.

10.8 SETTING THE SCENE

It is important that the training session has an introduction, a middle and an end. The introduction is about setting people at ease and reminding them why they are here. It may be that the group of trainees are diverse in some way, in that they come from different departments, different internal workplaces, different sites within the organisation, or even from different organisations or countries. In these circumstances it may be necessary to have some sort of ice-breaking activity to get people used to working together or involved. Some ice-breaker activities can be found in the Resources section of this book. The ice breaker is intended to break down barriers and get people thinking. One of the multi-national companies that I worked for insisted that every meeting (not just safety meetings) started with a safety message. Any member of the meeting could volunteer to talk about a safety message or event that they had experienced or heard about recently.

Even if the trainees already know each other, it is still necessary to do an introduction and put the training in context. The introduction needs initially to deal with the administration/hygiene factors and limitations of the training. This might include such topics as:

- Reminding about the start and finish times
- Arrangements for breaks and lunch
- Location of toilet facilities
- Provision and explanation of handouts
- Arrangements for validation (will there be a test at the end?)
- Request for everyone to get involved
- Fire/emergency evacuation arrangements
- Mobile phone policy
- Arrangements for updating training records
- Arrangements for feedback about the training session

It is helpful, if once the preliminaries are dealt with, that some sort of thought-provoking activity is used to set the scene. I find that most people consider themselves to be safety conscious and work in a safe way. To test this theory, I sometimes use the following exercise to see how safe we really are:

1. Ask everyone to stand up.
2. Ask people to stay standing up if they think that health and safety is important (hopefully everyone will stay standing! – but not always! If someone sits down at this stage ask them why).
3. Tell people that the most common cause of fatal accidents in the home is fire – so remain standing if they have a fire extinguisher at home.
4. Ask those still standing "What is it about fire in the home that is the most common cause of death – is it the flames or the smoke?"
5. Smoke is the main killer. Stay standing up if you have smoke detectors at home.
6. Stay standing up if the batteries in your smoke detectors have been changed in the last two years.
7. If some people are still standing up, then ask the final question. "Stay standing up if you have thought about how to escape from the upstairs floor of your house in the event that the staircase is compromised by fire, and that you have shared this escape plan with the remainder of the house occupants".

Despite the fact that that nearly everyone believes in behaving safely, it is surprising how few people are left standing at the end of this exercise. The point of this exercise is not to be clever, but just to get people to recognise that we don't always behave safely at home with our loved ones. It is not surprising if we take shortcuts at home, we are likely to sometimes behave unsafely at work.

Another very thought-provoking way to begin a safety training session is to use the following poem by Don Merrell, which is reprinted here with his permission. The poem can either be read out by the tutor, or with everyone reciting it aloud. The impact of this poem can be increased by reading it again at the end of the training session.

I Chose to Look the Other Way – I could have saved a life that day
By Don Merrell

I could have saved a life that day,
But I chose to look the other way.
It wasn't that I didn't care,
I had the time, and I was there

But I didn't want to seem a fool,
Or argue over a safety rule.
I knew he'd done the job before,
If I spoke up, he might get sore.

The chances didn't seem that bad,
I'd done the same, He knew I had.
So, I shook my head and walked on by,
He knew the risks as well as I.
He took the chance, I closed an eye,
And with that act, I let him die.

I could have saved a life that day,
But I chose to look the other way.
Now every time I see his wife,
I'll know, I should have saved his life.

That guilt is something I must bear,
But it isn't something you need share.
If you see a risk that others take,
That puts their health or life at stake.

The question asked, or thing you say,
Could help them live another day.
If you see a risk and walk away,
Then hope you never have to say,
I could have saved a life that day,
But I chose, to look the other way.

repoduced by courtesy of donmerrell@hotmail.com

10.9 PROMOTING DISCUSSION

To get discussion going, the tutor should always use "open" questions. An open-ended question is a question that allows the trainee to express himself or herself freely on the given topic and which requires a thoughtful response. Closed questions are those which elicit a "yes" or "no" answer. These answers do not indicate that the trainee is thinking about the subject in depth, and are more of a test of knowledge than understanding. The difference between closed and open questions is demonstrated by asking the same question in different ways:

Closed Question	Open Question
"Did you know... . ... .?"	"How do you know that... . ... .?"
Answer: "Yes" or "No"	Answer: An explanation is given

The important words to convert a closed question to an open one are to use: "How", "What" or "Why". The sort of phrases that can be used to open up a discussion might be:

- "About" Questions — "Tell me about an unsafe situation that you experienced recently"
- Hypothetical Questions — " What would you do if your workmate told you that something was unsafe?"
- Statements — "Susan, you look unsure about that. What do you think about it?"
- Silence — "... .... "

Silence is never as long as it seems! It also allows the tutor to listen. It is very difficult to speak and listen at the same time. We were born with one mouth but two ears, so we should be listening more than we are talking. I like to remember that the word "LISTEN" is an anagram of "SILENT".

The response to trainees' questions is just as important. It is important to make them feel valued. Compliment the trainees on a great question and do not put anyone down because that will result in them stop asking questions. When responding to questions, avoid compliments followed by the word "but" as that implies that you are lifting them up only to knock them down again! Also avoid using leading questions such as "Don't you think that it would be better to... . ... .?"

Once discussion begins, the role of the tutor is to steer the direction of the discussion.

Moving the discussion on can be helped by the use of the 4B's

1. Banter – Friendly light-hearted chat that creates a non-threatening atmosphere
2. Build – Take ideas that the trainees suggest and develop them to suit the session aim
3. Boost – Support the contributions from quiet/shy participants
4. Block – Interrupt disruptive participants by asking what everyone else thinks

Engagement can also be aided by small verbal interjections or gestures by the tutor:

Verbal	*Gestures*
I see	Smile
Ah ha	Nod
That's interesting	Maintain eye contact
Go on ...	Lean or move forward
Tell me more about that ...	Raise eyebrows
OK, so why don't you	Frown

The importance of listening is fundamental to getting engagement and trust. The concept of Active Listening is a technique that is sometimes used in training. It requires the listener to fully concentrate, understand, respond and then remember what has been said. The application of active listening requires the tutor to:

- Pay attention.
- Show that you're listening.
- Provide feedback.
- Defer judgment.
- Respond appropriately.

Maintain a list on a flip chart or whiteboard of questions that you will need to come back to before the end of the training so that trainees don't think that you have avoided answering their questions.

During discussion keep an active watch on the verbal and non-verbal signals from the trainees. Watch for signs of confusion, boredom and ensure that you respond to that feedback.

Every discussion needs to have a clear training objective in the mind of the tutor, so that the conclusions support the training and do not contradict it. When teaching accident investigation, there is a particular point that I usually want to make about who is likely to be accountable for the root cause of injuries. I begin by asking the group what are the sort of general recommendations that tend to come out of most accident investigations? The group brainstorms ideas onto a flip chart and tend to come up with a list which identifies such things as shown in Fig 10.2:

The group are then asked whether it is mainly the individual or management in the organisation who is responsible for each of the outcomes on the flipchart list? This is particularly interesting when done with a group of managers. Almost without fail, the groups identify that the responsibilities for the provision of up-to-date procedures, training, the specification and supply of PPE, the enforcement of safety standards and the provision of appropriate information all lie primarily with management.

Common Incident Outcomes

- Revise procedures
- Training
- Better PPE
- Enforce standards
- More information
- Custom & Practice

Figure 10.2 Common incident outcomes.

The group is then asked, "who therefore is most likely to be accountable for failures leading to injuries & incidents?" The group readily conclude that the root cause often resides largely with management. By using this targeted approach, the group have concluded by themselves the point that I wanted them to appreciate; that the cause of injuries is not solely the responsibility of the injured person. They remember that much more clearly than if I had just told them!

10.10 TEMPO

The speed of a presentation or training activity will be dictated by a number of things:

- The breadth of the scope of the training. - The more that there is to cover, the faster the training may need to be.
- The starting knowledge of the trainees. - Can you assume that they know the basics, or do you need to start from first principles?

- The quantity of training material that you need to get through.
- The number and complexity of practical activities.
- The more questions that you have to deal with, the more likely you are to run out of time and therefore to increase speed.
- Getting trainees to develop answers for themselves (i.e. workshop approach) is very effective but takes time.

Generally, if the trainees' level of starting knowledge is low or the topic is seen to be complex then the training tempo will need to be slower. From time to time, check with the trainees by asking "Is that OK or am I going too fast for you?"

In terms of the tutor's tempo, then if you are snappy, you may come over as authoritative or like a schoolmaster. This is fine in a large group such as a lecture hall where the audience accept you as the "Oracle" but in smaller workplace groups it can be intimidating or it might suggest that you have lost control of the timing.

Slower delivery on the other hand tends to suggest a more leisurely and consultative approach. In this situation the tutor is more of a facilitator than an instructor. This tempo is more appropriate to practical learning but might also be condescending for some more experienced trainees. I recommend a hybrid approach of using both fast and slower delivery to ensure that trainees get time to understand the fundamentals but that other parts can be gone over quite quickly. Maintaining this balance is important because it is important that the tutor goes slow enough to ensure understanding, but fast enough to cover everything that is needed in the time available. Varying the tempo throughout a training session will make it more interesting and memorable and reduces the chance of boredom setting in.

Controlling tempo is not just about the speed at which the tutor talks. A slower tempo can be achieved by the use of more anecdotes, case studies, examples and discussion. This approach is more relevant when the training involves the transfer of skills and practical applications. A faster and more authoritative delivery can be more exciting and stimulating. A fast tempo may be more appropriate when communicating knowledge or information. This might be in a situation where the training relates to the introduction of new legislation or local safety procedures. The accomplished tutor is likely to use different tempos throughout the training as a means of maintaining interest. The main thing about tempo is to ensure that the tutor is in control of the speed of delivery and that it does not arise by default because you have run out of time!

10.11 PUBLIC SPEAKING

Even the most experienced actors will tell you that they are nervous before going on stage. It is the same with making a health and safety presentation.

It doesn't matter how experienced you are, it is natural to be nervous. In fact, it is not only natural, it is important, because it shows that you really want your presentation to go well. That in turn shows that you care about both your subject and getting your message over, in a way that the trainees will understand and remember. The best way to control your nerves is by being well prepared and practised. One of the problems with modern IT systems is that they are fickle. It seems to me that not all computers talk to all LCD projectors and some permanently installed projection systems use VGA connections, whereas others use HMDI or USB. What is more, some systems time out after a while to conserve energy and require a password to reset them. I was doing a presentation in Switzerland for a client and noticed when I was testing my presentation just before the session started, that there was several seconds delay between clicking the mouse and the slide advancing. My host explained that the reason for that was that their IT system used dumb terminals located off their mainframe computer, so that when the mouse was clicked a message was sent to the mainframe and back. It turns out that the mainframe computer was located in Belgium! Things like that can upset your presentation and confidence if you do not know about them. So, make sure you get the chance to check that everything works before your presentation starts. In particular, if there is a sound system or you are using video clips with audio, make sure that the volume control is plugged in and switched on. It spoils your presentation if you casually go into a video only to find that there is no sound or it is so quiet that people at the back cannot hear it! It looks really unprofessional to be fretting about how to get sound to a video that is already running.

The well-prepared presenter will have a series of speaker notes, however, it is poor practice to be continually referring to notes or reading the presentation, as this detracts from the ability to pick up the signals from the audience. Not only that, but it tends to convey the message to the participants that you are unsure about your subject, and they need to be sure that you are competent to train them. If you need to have speaker notes immediately to hand, have short prompt words and bullet points on a small pack of cards that you can inadvertently hold in your hand. Remember not to write too small or to try and cram too much on each card. To avoid dropping the cards and changing the order, punch a small hole in one corner and string them together so that if they do get dropped you can easily find your place again.

Many inexperienced or ill-prepared speakers tend to read the text from the slide showing on the screen behind them. This can be a bit insulting to the trainees, as you may be suggesting that they are incapable of reading it for themselves! The content of the slide should be a précis of what you are saying, so use the slide as a trigger for what you want to say, but make sure that you embellish on it.

When speaking or gesturing, emphasise or exaggerate important things to make the point. Be prepared to project your voice more than you would in a normal conversation and enunciate clearly. A good presentation is a saw-tooth of highs and lows. The presenter needs to create the highs by varying voice pitch and volume. The most boring speeches are often those that are delivered with a monotone. If you are not sure about your delivery style, then video record your presentation and play it back so that you can recognise and amend your voice inflections. The highs in any presentation will not just come from the way in which you deliver the spoken word. Activities and exercises that get the trainees involved will almost automatically raise their level of interest and engagement. We need to treat the total presentation like telling a story. If you think about a good book or film that you have seen and ask yourself why you enjoyed or remembered it, then we invariably find that it has a saw tooth plot. It will have a clear beginning, middle and end. The beginning will set out the background. There will be a cadence of highs and lows throughout the middle of the story as the hero encounters crises and then overcomes each one. And finally at the end, the hero lives happily ever after. The memorable presentation needs to follow a similar storyline. We start with a beginning that explains why we are here and where we want to get to. The middle section should be peppered with real-life examples and exercises that vary the tempo and support the point that is being made. Finally comes the "denouement", which is the final part of the presentation, when the conclusion is decided or made clear.

A good presentation will be like a yacht race. It may not be possible to go directly to the finishing line, if the wind changes the yacht needs to tack sideways or into a headwind. In the same way the presenter will meet challenges and questions during the presentation that need to be dealt with, but he or she still needs to get to the finish line in time.

Presenting and training, generally, are dependent on the tutor's skill of communication. Just because the tutor said something does not necessarily mean that what was said has been heard! There could be a number of reasons for this. It could be that the trainee misheard what was said, or perhaps his or her attention was distracted at the time. It may be that the tutor used technical words that the trainee did not understand, or it could be that the trainee was just bored or heaven forbid – asleep! As trainers we must remember that our objective is not just to get the trainees to hear what we say, we want them to go away and continually apply what we have said. To achieve this euphoric state of understanding there are several hurdles to overcome in any presentation or training situation. The tutor must recognise the Basic Laws of Communication:

Said is not yet heard

Heard is not yet understood

Understood is not yet accepted

Accepted is not yet applied

Applied is not yet maintained

It is only by repeated checking of the trainees understanding at each stage that we can be assured that on completion of the training and subsequent validation and monitoring that the training has been understood and is being applied in a consistent way in the workplace.

10.12 ATTENTION SPAN

Attention span is the amount of time we spend concentrating on a task before becoming distracted. Distraction occurs when attention is uncontrollably diverted by another activity or sensation. According to the Associated Press, the average attention span is now around 5 minutes long. Ten years ago it was 12 minutes. Twenty-five per cent of people periodically forget the names of close friends and relatives and 7% sometimes cannot even remember their own birthdays! Younger people have a shorter attention span than older age groups. The use of social media is re-wiring our brains. As trainers, we need to understand that maintaining trainee concentration is crucial to the effectiveness of our input. The tutor needs to think in advance about how to minimise distractions and maximise interest and mental stimulation. Remember to arrange the training location so that external distractions are kept to a minimum. This means making sure that trainees aren't spending all their time gazing out of the window or looking at their mobile phones (research shows that most office workers look at their emails about 30–40 times/hour – so get them to switch off their phones, turn them to silent or even better switch off the router!) If training is going on in the workplace, try and ensure that there are no other activities going on that might distract the trainee.

In classroom style training, addressing attention spans is particularly crucial to success. The younger the trainees are, the shorter is their attention span. When using "all talk" lecture type of training it is highly lightly that concentration will have waned in less than 10 minutes. The tutor needs to introduce variety to the training pace or style every 10 minutes or so to keep the trainees' attention. This means frequently changing the format of the training. This could be to introduce an exercise, activity, video clip, physical demonstration, competition or quiz to move the training style away from just talking and showing slides. Unfortunately, it is the planning of these additional activities that is quite time consuming.

In the work environment most shop floor workers will not be used to sitting in one place for long periods of time. Training needs to address these norms and must attempt to avoid comparisons with being back at school under the tutelage of a cane-bending disciplinarian. I take every opportunity to get trainees moving around. Using such training aids as photohazards searches, flashcards and jigsaws (see Resources sections A, B and E) all get the trainees engaged in learning activities within the training room, but with the need to stand up and move around. Sometimes a simple ice breaker exercise which is or is not directly connected to the training subject may be necessary to get blood circulation going again or raise the levels of interest.

Always remember that if people are enjoying themselves then you will have their attention. Safety is a serious subject but that does not mean that there should not be times for being light-hearted. Video clips from TV "blooper" shows can be used very effectively to demonstrate the consequences of reckless actions. However, be careful not just to use this type of video clip to get cheap laughs – there must always be a learning point that comes out of any photograph or video clip that you use. Just one word of warning – the objective of a good training session is to engage people and make it fun. I have seen some trainers who are so keen to make the session "fun" that their objective becomes just "being entertaining" and they lose sight of the importance of the underlying message.

It is relatively easy to liven up even the most dreary of subjects by the use of "rewards"! I often give out chocolates in response to people calling out the right answers to the questions that I pose to them. This use of rewards engenders a feeling of competitiveness which raises everyone's attention and makes the session fun. Again, be careful not to make the reward of any real value or to overuse the "rewards" such that it becomes an expectation. Reward incentives are just one small part of the tutor's armoury. A particularly good use of rewards is if trainees are expected to come and present some pre-work that was done in advance of the training. I use this as a means of reinforcing the point that we all have something to contribute to health and safety learning – it is not the sole preserve of the tutor. On a five-day International Health and Safety Workshops that I run, I ask each delegate to come prepared to talk for 5 minutes about a health and safety situation that they have experienced previously. This could be an accident or some idea that they have used to improve health and safety in their workplace. The presentations are made a bit competitive by the fact that the delegates select the best presentation themselves and then a small engraved trophy is presented at the end of the course to the person that they chose as the winner. A fun element is also added during the presentations. The challenge for the tutor is to keep the presentations to a limit of 5 minutes. The delegates are told in advance that a timer will be run on a computer from the start of each presentation. The timer operates silently for the first 6 minutes (allowing a 1-minute over-run) and then Mr Tchaikovsky and his band come in playing the 1812 overture at around 90 decibels. The delegate quickly realises that he cannot carry on – much to everyone else's delight.

The use of chocolates as rewards has a double benefit. Sugar is the brain's preferred fuel source. We are not talking about refined "table" sugar but glucose which the body processes from the sugar and carbohydrates that we consume as foods. This sugar enhances alertness and therefore if your training course is providing refreshments, then fruit juices such as orange juice can offer a short-term boost to memory thinking and mental agility. However, too much glucose can cause the memory to be impaired. Caffeine also acts as a central nervous stimulant and can help mood and make people more productive, alert, energetic and help with short-term memory.

It is worth thinking about the foods and beverages that you order for the breaks during training. Too much food at lunchtime is likely to make people drowsy. However, according to Chris Bailey, nine brain foods that will improve focus and concentration are:

- Blueberries
- Water
- Tea
- Avocados
- Leafy green vegetables
- Fatty fish
- Dark Chocolate
- Flax seeds
- Nuts

On the other hand, you may find that hard-working practical staff find that a hot bacon butty or piece of pizza puts them in a better mood!

10.13 EMPHASIS

We have learned previously of the importance of varying speech pace, volume and intonation in order to avoid monotony, which inevitably leads to boredom. It is essential that the emphasis comes at the right time, so that the trainees remember the key points of your presentation. A simple but effective way of reinforcing a particular point is to simply repeat it. To be effective the repeated message should be brief and to the point. This is a common technique used by politicians to get us to remember a particular part of their message. Having made your point, wait for a couple of seconds and repeat the message, but this time a little more slowly. Bringing the same point up on a separate visual image or writing it on a whiteboard or flip chart also increases the impact. The value of using the handwritten message on a flip chart is that it remains visible for much longer than a PowerPoint slide, which gets overwritten with the next click of the mouse.

Practical demonstrations can also be used to emphasise messages. When wanting trainees to remember the importance of taking a balanced approach to the different aspects of safety, I use a homemade model of a three-legged stool. Each leg represents the three "P"s – that is People, Procedures and Plant. The point is that each of these has to be addressed equally to keep the stool legs equal. If we forget about the importance of people and focus only on procedures, I demonstrate that the stool legs are of different lengths and the stool falls over and your focus on safety collapses. The practical use of this demonstration requires that part of one of the legs of the stool is easily detachable as shown in Fig 10.3.

Figure 10.3 The three-legged stool.

10.14 USE OF POINTERS, LASERS AND PRESENTERS

During any presentation it will be necessary for the tutor to want to refer to or draw attention to a particular aspect of their presentation slide. This is often done by the use of some sort of "pointer". This pointer may just be the tutor's hand, or a long pole, a laser light beam or the use of the computer's cursor. The use of the correct type of pointer is not just a matter of personal preference. It will depend on the size of the location and the technology that is available. In an ideal world the tutor's hand is the best and most natural way of highlighting a particular point because the tutor will be able to face the audience. This is only possible if the display screen is of a size where the tutor can reach to highlight the points on the screen. If the presentation is using a table-mounted LCD projector, there are two big drawbacks from this method of manual pointing – the tutor may be dazzled by the projector beam and may cast a shadow over large parts of the screen. To overcome this problem, if the presenter is standing beside the screen or VDU, the manual method can be improved by the use of pointing stick or a telescopic pointer (available from most office equipment suppliers) as this allows the presenter to stand to one side and not interrupt the projection. The use of a hand or finger to highlight points can go wrong if the screen is touch sensitive, as any contact with the screen has the same effect as a mouse click and will advance the slide or cause some other unwanted action.

If it is a very large or high screen, then laser pointers are often used. Many presenter remote controls have an integrated laser pointer which can project a red laser spot onto the screen from a considerable distance. It is worth noting that if used with a visual display screen with a non-reflective surface, the laser spot may not show up. There are some limitations with the use of laser pointers. Very often the tutor is at some distance from the spot and so the trainees cannot focus on the tutor and the spot on the screen at the same time. This is not a problem with a large audience but does affect the impact of the point in a smaller training room. The second thing to be aware of is that there is a tendency for the tutor to misuse the laser by either moving it around very quickly or flashing it on and off very quickly such that the viewer cannot register exactly where the laser is pointing. Finally, it is worth recognising that the further away from the screen the laser pointer is held, the more difficult it is to hold the laser spot steady and precisely position it on the screen. Many presenters have an annoying tendency to circle or wobble the laser around while talking, which makes it very difficult for the audience to focus on. The laser pointer is useful when pointing out details on an engineering drawing, or photograph, but it is always preferable to use text animation rather than a laser when highlighting wording.

The last type of pointer that is increasingly being used is where the presentation uses a large computer visual display screen. Often the computer cursor is used as a pointer. This is fine in a business meeting situation, but in a training context this often results in the tutor running the whole presentation from a seated position. This unintended consequence means that a lot of the energy and dynamism of the presentation is lost, and that the tutor will not be able to move about and look trainees in the eye. Use of the cursor as a pointer should be used very sparingly.

It is advisable for the tutor to have a personal "presenter" in his training kit. A "presenter" is a special form of remote mouse that runs on a wifi link from the computer's USB port. The presenter allows the tutor to advance, retard, blank out or laser point the slides. The reason for having a personal

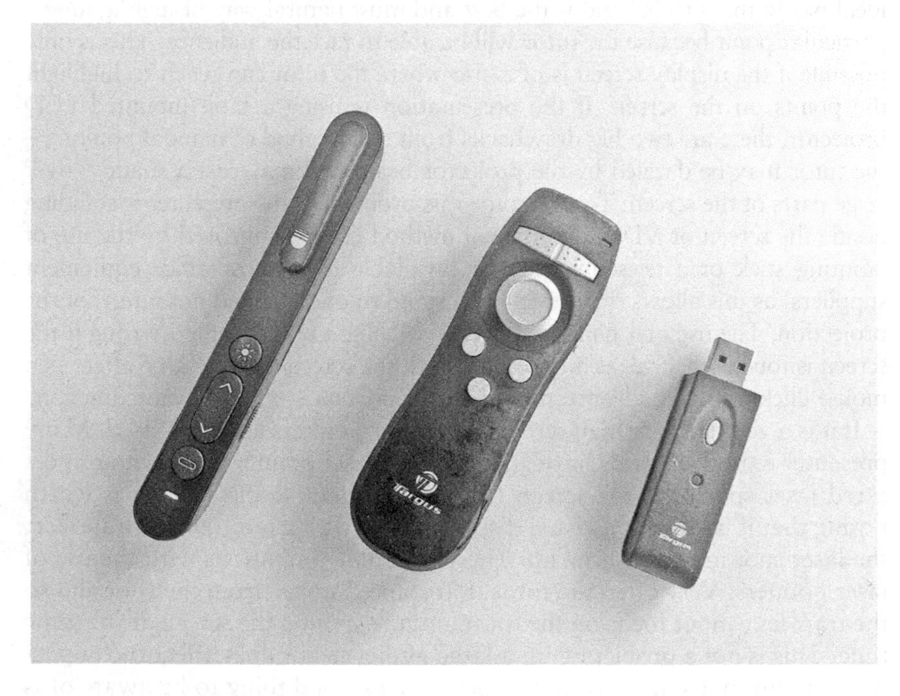

Figure 10.4 Types of presenter.

presenter is that some training facilities do not have remote mouses available and only supply a wired mouse. This means that the presenter is tethered to the computer throughout the session with all the accompanying lack of involvement (Fig 10.4).

10.15 PILOT EVENTS

It is important to practise your delivery skills. If the training session is the first of what are expected to be many similar events, then it is advisable to run a pilot event. The pilot event is like a dress rehearsal for the training session. It is done after all presentational material and exercises/activities have been developed to the level that the tutor is satisfied that they meet the training specification and timescales. The pilot session is then run with a group of carefully selected individuals who understand that their role is not only to take part in the training, but then subsequently provide constructive feedback and criticism. It is important that the feedback from the pilot event is taken into account and that the training material is amended as required. When planning the training programme, allow sufficient time between the pilot session and the next session to allow for any modification of the training materials.

10.16 WHEN THINGS GO WRONG

Things can go wrong for even the most experienced trainer. As Murphy's Law tells us, "If anything can go wrong, it will!". The corollary to this is to be prepared. If you have practised, made provision for accessing spare projectors and other training hardware and have several copies of your training material available, the chances of unplanned things going wrong are minimised, but not totally eliminated. Most of us have had sessions that we would rather not repeat but the important thing is to learn from those, change your approach and move on. Most importantly do not allow one bad experience put you off. It may be a glass half empty to you but a glass half full to the trainee. Remember like an actor, only you know exactly what should be happening, so often the audience may not notice minor deviations from your plan. The key point is to keep going and don't let some small hiccup put you off your stride.

How do you know when things go well? You'll know without needing to be told. A successful training session will leave you with a huge buzz and feeling of satisfaction!

In summary, when delivering a health and safety training presentation, remember the key points:

The Art of Presentation Delivery Checklist

IT'S ALL ABOUT GIVING A PERFORMANCE!

1. Remember EDGE – your job is to:

   Educate

   Demonstrate

   Guide

   Excite

2. There are three phases in any presentation which are usually summarised as:

   - Tell them what you are going to tell them (Explain why we are here)
   - Tell them (The main learning)
   - Tell them what you told them (Summarise the key messages)

3. Address trainees by name. If necessary use name cards or name badges, but always address trainees by their known name.
4. Dress Code – dress in a way that makes the trainees feel comfortable and demonstrate that you have empathy with them.
5. Set the Scene – describe the hygiene/programme/lunch/safety arrangements.
6. Mannerisms – Think about what irritating mannerisms you might display and try and control them. If necessary video record your draft presentation and critique it yourself.
7. Make sure that the presentational equipment works.
8. Practice your presentation.
9. Anticipate obvious questions.
10. Promote discussion – involve everyone.
11. Vary the pace and tempo of the presentation – A slower tempo can be achieved by the use of more anecdotes, case studies, examples and discussion.
12. Do not just read out your notes or slides. Engage the audience – to have impact you must make the training interesting and fun to hold the trainees attention.
13. Deliver a performance.
14. When people are smiling, you can teach them anything.
15. Be enthusiastic.
16. Repeat: Recap: Review
17. Manage your time to avoid an overrun.
18. Keep it simple and clear.
19. It is quite normal to be nervous.

Chapter 11

Running exercises and activities – Making it fun!

Possibly the most important part of any training is where the trainees practise their learning by using exercises, activities or practical working. For most trainees this is likely to be the most interesting and valuable part of converting their newly acquired learning into further knowledge or a usable skill. The design of this practical work requires the same amount of thought, attention and preparation as the trainer would normally put into preparing the training presentation and handout materials. There is often a tendency to treat these crucial exercises as some sort of afterthought. If these activities are to have real value, then they need to be clearly understood, relevant, interesting and capable of being conducted within the time available for the overall training.

11.1 HOW TO PLAN THE EXERCISE

When planning an exercise, the tutor needs to consider what the objectives are:

- What skill is the exercise/practical work trying to reinforce? Exercises can be used either to practice a new practical skill, to develop a solution to a problem or practising the application of a new procedure. If the need is to practise a skill, such as the use of lifesaving defibrillators, then the exercise will need to have spare fully charged defibrillators and a warm carpeted room in which to practise. If the need is to practise a new procedure, such as confined space entry preparation, it is likely that the exercise will need to be done on a piece of out-of-use or redundant equipment away from the training room. However, if the objective of the exercise is for the trainees to do some original brainstorming exercise or develop a solution to a problem, more preparation needs to be put into posing the right questions, rather than arranging for specialised hardware or equipment is available.

DOI: 10.1201/9781003342779-12

- How will the exercise build on the new learning?
  Unless the purpose of the exercise is as an "ice breaker", the proposed activity must be directly relevant to the subject of the training and related to the learning that immediately precedes it. The practical work should reinforce the learning message and give an opportunity to test the extent of individual's understanding.
- Will the exercise be done individually or in groups?
  Group or team working allows individuals to learn from one another and provides opportunities for inter-personal discussion. This can make group activities more interesting and less individually threatening to the trainees. However, it must be remembered that it is possible for individuals to hide their lack of understanding within a group and apparently successfully complete the group activity with the minimum of involvement and contribution.
  There are certain practical exercises, such as breathing apparatus training which will always require individual practice and cannot be done in groups.
- What equipment or tools will be required to carry out the exercise?
  If the activity or practical work requires special equipment, tools or instruments, then a check needs to be made to ensure that enough of these are available on the day of the training and if necessary that they have been recently calibrated. If using dedicated training equipment, make sure that all the examples are of the same model/type and that these are the same version as the trainees will need to use once fully trained and competent. If trainees are required to use their own tools or equipment during the practical work, make sure that they are aware of this before they attend the training.
  When the exercise requires the use of portable electrical equipment, such as computers, do ensure that these have a valid portable electrical appliance test. If you ask the trainees to use electrical equipment with is either not tested or out-with its test date, your health and safety credibility will be undermined.
- Where will the exercise be done?
  The location of the exercise can have a significant effect on how well it is received and also how long the exercise will take. If the exercise is a simple tabletop activity, like the flash card exercise in the resource section (resource E), then this can easily be run in the training room environment, whereas if the exercise involves practising isolating a piece of equipment in readiness for maintenance, that is likely to need to be done in the workplace. The location of the practical exercise will have a significant effect on the time that needs to be allocated, especially if the exercise requires the trainees to go to another location and perhaps have to receive special permits or obtain and wear special items of personal protective equipment.

If the practical work is to be run in a public area, like the workplace or a workshop, try and arrange things so that it does not clash with other events that might distract the trainees' attention. Even other people passing for a chat or wanting to know what is going on, will all disrupt the training and add to the time commitment.

- How long will it take?

  The total time allowed for the activity must factor in not only the actual time for the practical work, but also travelling, preparation and feedback times. Depending on how far away the activity is to be run, it can often be that the actual activity time is only half of the total elapsed time in the training programme. The total time for an activity needs to allow for the combination of the times to:

  i. Brief the trainees on what has to be done
  ii. Travel to the location or break-out room
  iii. Carry out the exercise/practical work
  iv. Return to main training room
  v. Feedback on outcome of exercise by both the trainees and the tutor

  It is surprising how time for exercises adds up. There are not many exercises that can be completed in much less than half an hour. Even if the exercise is being done in a near-by break-out room, it is easy to lose 5 minutes before and after the exercise as people get distracted or need to answer a call of nature. The initial briefing of the task to be done, needs to be kept as clear and simple as possible and should always be reinforced with a short, written brief but even then, this is likely to take 5 minutes. The brief should make clear not only what needs to be done, but also what sort of feedback is required by the group or individual. If the feedback requires to be on a flip chart or PowerPoint slide, then the time allowed for preparing this can be quite significant, and this is then followed by the actual time required to stand up and present the feedback and for the tutor to reply with his or her comments. If there are several syndicate groups independently carrying out the same exercise, the groups' feedback is likely to be very similar. In this situation, to avoid everyone becoming bored, hearing the same comments repeated several times, the overall feedback time can be reduced substantially by reporting by exception. This means that the second and subsequent groups to feedback do not repeat comments made by the first group in their feedback but only add additional points that have not already been covered. Alternatively, to make the feedback very brief, the tutor can run the feedback in the plenary session and just invite trainees to call out what they found during the exercise.

The key thing for the designer of the training exercise to recognise is that it is very difficult to run a significant exercise away from the main training room in much less than a total elapsed time of an hour. In many cases, it can be much more than that. It is quite possible that the design of the practical work/exercises is the factor that really dictates the overall duration of the training.

- Will additional tutors/mentors be required?
When complex or detailed practical activities are used within a training environment, it is normal practice to break up the trainees into smaller syndicate groups of between four and six people each. If the exercise or practical work is straightforward and does not involve risk to the individual trainees, then one tutor can oversee several syndicate groups. This overseeing role should ensure that:
  i. The group have understood their brief.
  ii. They have whatever support equipment that they require.
  iii. They converge their work or thinking towards the final objective.
  iv. They are given a 5-minute warning of when they need to finish.
  v. They prepare the appropriate feedback and appoint a speaker.

If the practical work involves complex arrangements, access to hazardous areas or equipment or the need for detailed further instruction, then it is likely that additional tutors or instructors will be required. When these extra tutors are involved, they must also be well briefed about the task and their role, and this should be supported by a written tutor's brief.

- What is the end point of the exercise/practical work?
The final end point of the exercise or practical experience must be clearly defined and achievable. It might be that the objective for a first aid trainee is to be able to perform cardiopulmonary resuscitation (CPR) or that the syndicate group jointly complete a risk assessment. The tutor must monitor the progress of the exercise and, if necessary, intervene to provide guidance to ensure that the exercise is completed on time. One of the challenges in any multi-group exercise is to ensure that all the groups finish at about the same time. Otherwise, if one group is tardy, the other groups will be left wondering what to do. It is in this type of situation that attention wanes and some people may start to get bored.

A wide range of ideas for practical work and exercises is found in the training resources section at the back of this book.

Running Exercises and Activities Checklist

1. What is the purpose of the exercise?
2. What earlier learning is it intended to develop?
3. Is the exercise to be done individually or in groups?
4. What equipment or tools will be required?
5. Where will the exercise be run?
6. How long will the exercise take, allowing for:
   - Briefing the trainees on what has to be done
   - Traveling to the location or break-out room
   - Conducting the exercise/practical work
   - Returning to main training room
   - Feedback on outcome of exercise by both the trainees and the tutor
7. Are additional tutors/mentors required?
8. Is the end point of the exercise clear?

Chapter 12

Validation and competence – Checking that they have understood

The law generally requires that people employed at work must be competent. Being trained is not the same as being competent. In addition to being trained, competence requires that you can safely and consistently apply that training in the work situation. When learning to drive, we are not permitted to drive alone on the road once we have completed a course of driving lessons. Before we can venture out onto the road as a new driver, we must pass a driving competence test. It is the same situation at work. Before an employee is deemed to be competent, it is necessary that some method of competence checking is used. This process is called "validation" and is usually applied twice. Once to validate the training received and then again before the individual is confirmed as competent.

12.1 VALIDATION

Validation of training usually takes the form of a simple test. Often this is a written questionnaire, that the trainee is asked to complete at the end of the training. The problem with this approach is that the validation is only as good as the questions in the questionnaire. It is also quite common to use "multiple choice" questions. In one training event that I ran some years ago, relating to confined space entry, one of the trainees got several of the multiple-choice questions wrong. There were four options for each question, so he was given another chance to complete the validation questionnaire. At the second attempt he continued to get them wrong, so he was given another attempt which he also failed. At this stage, the only answers left were the correct ones that he still hadn't identified. It was pointless to ask him to do the questionnaire again as he clearly hadn't understood or registered that part of the training. However, it did enable me to recognise what part of the training he had not understood and to go back and repeat that on a one-to-one basis with him until he understood the message.

After the training validation, the trainee should be allowed to apply the newly learned skill in a controlled and monitored environment. This control

DOI: 10.1201/9781003342779-13

is usually achieved at work by putting the trainee to work alongside another worker who is competent in that subject. This approach is known as "mentoring". The mentor is there to monitor the trainee when he/she applies their new skill. The mentor is not there to do the work for the trainee, but will recognise and correct any errors or shortcuts that the trainee makes. The duration of the monitoring process will depend on how quickly the trainee is able to demonstrate the ability to apply their training consistently and safely. When the mentor is satisfied that the trainee is competent, there is usually a final validation carried out by a third person (often the trainee's supervisor or first-line manager). This final validation is often conducted by observation alone, and rarely includes any written assessment. The subject and date of the validation needs to be recorded as evidence of competence for that task/skill.

The need for robust and detailed validation is more important when carrying out task training than when carrying out awareness level training.

12.2 MENTORING AND THE ROAD TO COMPETENCE

In the work environment, most people prefer to learn by doing, rather than just by listening. Practical application is often seen to be very important. When it comes to Health and Safety training, many tutors make the mistake of talking too much and referring too much to legislative requirements. The problem for the trainer is that many people who are attending a health and safety training session will tend to see the messages as obvious and just common sense. They may think that it is demeaning to them and will very quickly switch off. It is true that health and safety is not exactly "rocket science" and we are dealing with concepts that might arise infrequently. The bored trainee will invariably say "that has never happened to me". This is because they usually consider themselves to be very safe workers – in the same way that none of us think that we will have a road traffic accident because we are such good drivers. Apparently, it is all the other drivers who are the problem!

The task of the health and safety tutor is to capture and retain the attention of the trainee. This sounds very simple, but it is in fact quite a challenge. Gary Player, the renowned professional golfer was commended after a series of tournament successes, on how lucky he was. He replied "It's amazing, but I find that the more that I practice, the luckier I get!" The key point in becoming competent is practice, practice and more practice. However, when practising skills at work, we must always recognise the possibility that things can go wrong. You wouldn't want a newly graduated dentist pulling out your teeth, unless you felt that he was under the supervision of a fully qualified and experienced dentist, just in case anything went wrong. It is the same in the work environment, once trained in the basic skills, the trainee should be supervised or mentored in the application of those skills until he or she is shown to be competent. It is interesting that we fully understand and apply this concept of mentoring at work in the situation of apprentice training. In this situation, the trainee apprentice will have received a year's off-the-job practical training in a specialised training centre and then the apprentice is introduced to the work environment progressively. Initially he or she will be given minor, low-risk tasks, and as they prove their ability at those, they will be given more and more responsibility and more and more technically challenging tasks. At first, this will involve high levels of supervision and mentoring, until eventually towards the end of the apprenticeship, that individual is competent enough to work with minimal supervision. This approach has been established as a very effective method of training young people, but somehow, we forget to apply the same principles to training later in life. So often we tend to adopt the "Sitting with Nellie" approach, where training is reliant on the trainee copying what a co-worker does and they pick up both Nellie's good and bad habits in a very unstructured and uncontrolled way.

Being a mentor to a trainee is a challenging but very rewarding role. Mentors are experienced and trusted advisers, who can offer guidance and help to the trainees to maximise their development. It also has benefits to the mentor, who often learns as much as the trainee during the process. In order to be an effective mentor, the individual should exhibit the following qualities:

- Be knowledgeable about the subject
- Be recognised as "competent"
- Know and understand the requirements
- Not inclined to take "short cuts"
- Be patient
- Be tolerant
- Be willing to help
- Have the time to help
- Have high personal standards in relation to the workplace
- Be willing to share his or her experience
- Be interested and willing to help others.

Being a mentor is tough, but can have great rewards for the mentor, when he or she sees the results of their work blossoming in the good performance of younger or less experienced colleagues. It should be seen as an accolade, in that only the best workers are seen to be suitable to take on this very important role.

In some organisations, this role is viewed as involving additional work and hence worthy of additional remuneration. It is rare to find that there is direct monetary reward for this type of additional responsibility, but many good mentors relish the opportunity to help others, and in my experience, it is one of the workplace skills which is highly desirable when it comes to workplace promotions – so often the mentors do eventually get their deserved rewards.

It should be remembered that the road to competence is not a single destination, where once you get there, you are competent for the rest of time. Competence is an ongoing journey in which we continually need to adapt to changes in either the workplace, the technology that we use or the standards and regulations that apply. We must treat achieving competence as part of our continuous skills development. As an example, when I started working, there were no computers and no internet. Those are now an absolutely essential part of nearly everyone's daily work. But personal skill development is not just about learning to do new things – we need to be competent in the application of those skills. Many years ago, while working in the steel industry, I was trained to mount big grinding wheels. This was an important task and if not done correctly and in a timely manner, a huge slab grinding machine would cease production. More importantly, in those days, when abrasive wheel technology was less well developed, if grinding wheels were incorrectly installed, they could disintegrate and were quite a common cause of workplace injury. I was trained to mount those wheels correctly and safely. At that time, I was given the skill training and subsequently, through regular practice, I became competent. It is now many years since I used that skill. Abrasive wheel technology has moved forward, grinding machine designs have changed and I have forgotten some of the training that I was given. For those reasons, I am no longer competent to install grinding wheels, despite being competent in the past and holding a certificate of competence. Competence comes from the regular application of a skill leading to gaining experience, and this together with the ongoing learning that comes from your own and other people's experiences in relation to the application of that skill, leads to competence. If you fail to keep up the application of your skill, or fail to adapt to new technology or learning in relation to that skill, then your competence will inevitably decline. It is a management responsibility to monitor competence and identify when there are shortcomings which may require re-training. This is not as easy as it sounds. When recruiting new staff, we often ask to see evidence of educational achievement or previous training. What is much more difficult during a recruitment process is to be assured of an individual's competence. This usually needs to come from some practical testing of that skill or period of testing before the employment is made permanent, or through taking up references and talking with previous employers. The construction industry has become particularly good at this, because of their very transient workforce. They have developed what are called "passports" which allow different employers in that industry to have confidence in the training and competence of people that they are taking on. They also have industry wide systems for establishing competence and operating "licences" for the common types of mobile plant that are used by that industry. Unfortunately, this approach is not widely copied in other industries.

The important message from this chapter is that training alone is not enough – we need to be able to assure competence.

Validation and Competence Checklist

1. Is training followed by a period of mentored practical application of the new skill?
2. Ensure that both the training and subsequently competence are validated and recorded
3. How are mentors selected? Can they be shown to:
    - Be knowledgeable about the subject
    - Be recognised as "competent"
    - Know and understand the requirements
    - Not inclined to take "short cuts"
    - Be patient
    - Be tolerant
    - Be willing to help
    - Have the time to help
    - Have high personal standards in relation to the workplace
    - Be willing to share his or her experience
    - Be interested and willing to help others.
4. Are mentors recognised for their contribution?
5. Is there a system to ensure that the newly trained worker is assessed for competence before being allowed to apply the skill alone in a hazardous environment?

Chapter 13

Training administration

Sometimes the best standard of training can be affected by shortcomings in the administration. Often the tutor is not responsible for the administrative arrangements of providing a suitable location and notifying the trainees. It is important to check with the administrator that the arrangements are suitable and that any pre-course information has been sent out. I find that the most common cause of complaint about administration and location facilities is room temperature control – either too hot or too cold. In particular, rooms that are too cold can be very distracting for trainees and when they are too hot the trainees might have difficulty staying awake.

13.1 PRE-COURSE INFORMATION

Ensuring that trainees arrive on time, in the right place, and with any prerequisites, is essential to the smooth running of any training event. Both the trainee and his or her manager must be notified well in advance, as arrangements will need to be made to allow them to be released for that training. It is important that both the trainee and the manager are aware of the scope of the training and that the timing is appropriate, necessary and in line with the training needs of the individual trainee.

Pre-training information may need to cover some or all of the following:

a. Training topic
b. Brief outline of training/programme
c. Date and time of first training session
d. Duration of each session
e. Number of sessions
f. Location
g. Lead Tutor
h. Other trainees attending
i. Dress Code (Does it involve workplace training and protective equipment?)

DOI: 10.1201/9781003342779-14

j. What will be provided (notebooks/pens/handouts, etc.)
k. What will the trainee need to bring? (Are any special tools required or personal laptop computers?)
l. Does the trainee need to do any preparatory work?
m. Validation arrangements
n. Lunch and breaks (including any special dietary requirements)
o. Any special access pass/signing-in arrangements
p. Car parking facilities
q. Arrangements for handling incoming messages

If the training is taking part in an off-site facility like a commercial training centre or conference centre then it may also be necessary to notify the trainee about:

r. Accommodation arrangements (including directions to the training facility)
s. Travel arrangements
t. Expenses claim arrangements

It is usually helpful for all parties if the individual's boss and the trainee have discussed the aims of the training and why it is necessary, prior to the trainee receiving the pre-training information. It is also very helpful for the boss to have a feedback discussion with the trainee once the training is completed and before the monitoring stage starts (see Chapter 12). A short reminder about the upcoming training is often advisable a few days before the training starts just in case it has got forgotten.

13.2 EQUIPMENT LISTS AND TRAINERS KITS

It doesn't matter what sort of training is being done, it will require some sort of equipment, even if that is only pens and paper. My experience is that the more activities and demonstrations that the session has, the better the outcome, but the more complex that organising the session becomes. Complex sessions may be much more enjoyable and interactive but they also have more opportunities to go wrong. It is essential to develop an "Equipment List" to ensure that nothing gets forgotten or is left to chance. This list is a reminder to the trainer of all the equipment that is necessary to successfully run a particular training session. It is very easy to forget to bring the spare projector or the handouts for a particular exercise. The equipment list is particularly valuable when the session is one of a series of repeat training sessions.

a. Most training sessions will require presentational equipment such as:
  - Laptop, computer or tablet
  - Large VDU or LCD projector, screen and linking cables
  - Independent audio speakers
  - Flip chart stand/flipcharts/marker pens
  - Backup for presentation electronic files
  - Floor cable protectors or gaffer tape

b. The training room facilities will require
  - Enough adequate seating for all registered trainees
  - Sufficient space for the tutor's preferred layout
  - Suitable tables/writing surfaces
  - Pens and paper
  - Name cards
  - Soft drinks and disposable cups
  - Draughting tape/blutack for displaying flipcharts
  - Clock

c. The Presentational material will require
  - Enough handouts for all the trainees + tutors + one extra spare
  - Copies of all exercise materials and equipment
  - Attendance register and feedback forms
  - Rewards and incentives (chocolates?)
  - 4-hole punch
  - Room signs (to indicate the training session title outside the door)
  - Presentation files electronic backup copy

The following is an example of a bespoke equipment list for an auditing skills training event.

	Equipment List – Example	*Out*	*Return*
1	Laptop, computer/tablet + spare		
2	Presentation file backup memory stick		
3	LCD projector and remote control + access to spare		
4	Projection screen/Whitewall or VDU		
5	Extension speakers		
6	Extension cables and RCD (PAT tested)		
7	Cable tape (Gaffer Tape)		
8	Cable cover		
9	Cable tidy		
10	Socket 2-way adapter		
11	Spare connection cable box (VGA, HMDI, Audio Jacks, etc.)		
12	3 Flip chart stands with pads		

(Continued)

Equipment List – Example		*Out*	*Return*
13	Marker pens		
14	Drafting tape/Blutak		
15	Placename cards		
16	Course manuals/handouts		
17	Room signs		
18	Clock		
19	Stationary box		
20	Digital camera		
21	Fully charged digital camera batteries		
22	4 hole punch		
23	Briefcase		
24	Trolley/Sack barrow		
25	Rewards (mini chocolate bars)		
26	Bottled water and disposable tumblers		
27	Training attendance sheet		
28	Training feedback sheets		
29	Briefing note for exercise 1		
30	Materials for exercise 1		
31	Briefing note for exercise 2		
32	Materials for exercise 2		
33	Etc.		

The "Out" column is ticked to ensure that everything that is needed to run the training session is available and taken to the training location. It is important to record a tick in the "Return" column at the end of the training to make sure that everything is available when required for the next similar event. It is a good idea to keep those items that are unique to this particular training in a box file to minimise the amount of preparatory work needed for the next time the session is run. In particular, much work can be saved if a master copy of the handouts and exercise briefing notes are kept in this file. If electronic files are maintained, then it is a good idea only to keep the current versions of the files, otherwise Murphy's Law will apply again and inevitably you will arrive at one training session where you and the trainees are looking at different information.

13.3 STATIONARY BOX

One of the best ideas that I picked up from a mentor of mine is to have a stationary box. That is not a box that is immobile, but it is one which contains a selection of items from the office stationary cupboard. It doesn't matter what the training subject is or who is attending, there will always be a need for those critical odds and ends. I am always amazed when trainees turn up for training and they don't even have a pen or piece of paper with them. A lot of time can be lost while you or they go off in search of the missing items. By having a permanently stocked small box of essential stationary items the inevitable shortages can be pre-empted and swiftly dealt with (Fig 13.1).

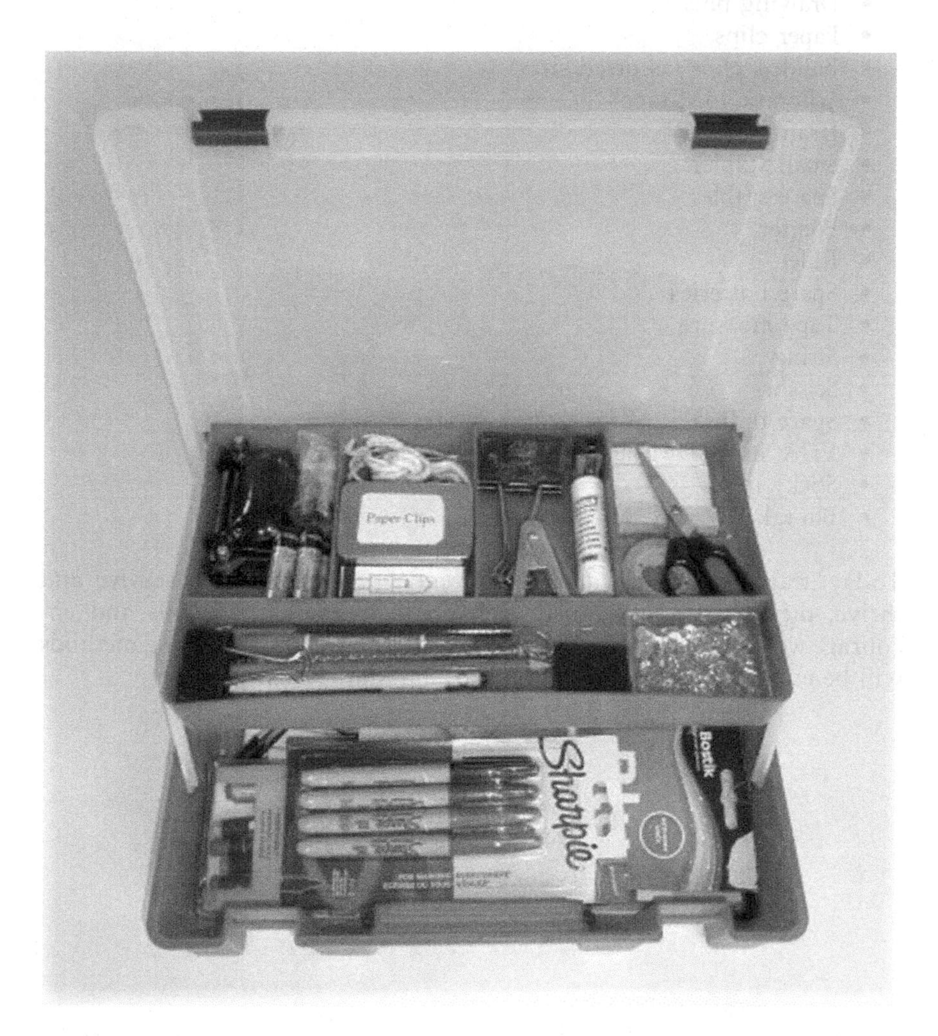

Figure 13.1 A typical portable stationary box.

The contents of the stationary box typically will include:

- Pens
- Pencils
- Erasers
- Small pencil sharpener
- Marker pens
- Highlighter pens
- Drawing pins
- Paper clips
- Bulldog clips (assorted sizes)
- Sellotape (3M tape)
- Drafting tape
- Small stapler
- Spare staples
- Pointer
- Ruler
- Spare batteries
- Tape measure
- String
- Scissors
- Spare bulbs
- VGA and HDMI lead
- Sticky notes
- Blu-tak

Today, training is rarely a one-off event. For companies to survive and thrive, they must continuously develop their employees' skills and encourage workplace learning. So, in most cases, a variety of training methods will be useful at some stage of the learning journey.

Training Administration Checklist

- Is the location suitable?
  - Size/capacity
  - Temperature warm/cool enough
- Have arrangements been made for:
  - Meals
  - Provision of refreshments
  - Provision of tools and equipment needed during the training event
- Has pre-training information been sent out?
  - Training topic
  - Brief outline of training/programme
  - Date and time of first training session
  - Duration of each session
  - Number of sessions
  - Location
  - Lead tutor information
  - Other trainees attending
  - Dress code (does it involve workplace training and protective equipment?)
  - What will be provided (notebooks/pens/handouts, etc.)?
  - What will the trainee need to bring? (Are any special tools required or personal laptop computers?)
  - Does the trainee need to do any preparatory work?
  - Validation arrangements
  - Lunch and breaks (including any special dietary requirements)
  - Any special access pass/signing-in arrangements
  - Car parking facilities
  - Arrangements for handling incoming messages
- Has trainer got a "Stationary box" containing foreseeable requirements such as pens, paper, etc.?
- Have trainees' individual bosses briefed them on what the aims of the training are?
- Are arrangements in place to mentor and validate the trainees after completion of the training?
- Have attendance lists been provided and training records updated?

Chapter 14

Evaluating the effectiveness of the training – Did it work?

Training only has a value if it results in an improvement in the trainee's skills or knowledge. Without some method of evaluating the success of the training, leaves us blind to whether or not the training has been understood and applied. There is a tendency for some trainees and tutors to want to make training assessment easier, particularly where pass rates are an important measure. The assessment must be conducive to the risk. A high-potential risk operation is likely to need a more extensive and thorough assessment than for a lower-risk facility. The important thing for employers is that they need to be able to show that someone is competent to do a particular task and a fancy new certificate is not necessarily going to provide that if the training assessment has been insufficiently rigorous.

14.1 THE FOUR LEVELS OF TRAINING ASSESSMENT

The four levels training assessment model, developed by Donald L. Kirkpatrick in the 1950s, remains widely used today. The model identifies that when evaluating the effectiveness of training it is necessary to consider: emotional reaction, achievement of objectives, behavioural changes and organizational impact.

a. Level 1 training evaluation – Emotional reaction
   This level of evaluation looks at the attitude of the trainee when the training is completed. It is what commonly happens at the end of a training event, where the trainees are asked to give their feedback in a simple questionnaire (see Chapter 12). Typically, the questionnaire will give a subjective assessment of the trainee's attitude to the tutor and the training. Usually this would cover style, content and hygiene factors such as the quality of the training room and the knowledge and effectiveness of the tutor.

   This type of survey is very easy to arrange and gives quick feedback to the tutor and management as to how much the trainees enjoyed the

DOI: 10.1201/9781003342779-15

session. Normally this type of feedback process will assess the level of the trainees' satisfaction with such things as the relevance of the training to their needs and the quality of the training provided. The results can be easily quantified as shown in the example below of a typical feedback sheet (Fig 14.1) and have the benefit of readily identifying whether successive training courses are improving or deteriorating. Understanding whether the trainces' satisfaction with the training and whether they have enjoyed the session is important, because if they haven't enjoyed it, it is very likely that they will have learned much.

Health & Safety Workshop– Delegate's Feedback

Delegate's Name (Optional):

	Relevance of Content					Quality of Presentation					Comments
	1	2	3	4	5	1	2	3	4	5	
Managing Safely											
Human factors											
Risk Management											
Learning from Accidents											
Management of Change											
Process Safety Management											
Safe Systems of Work											
Managing Health Risks											
Sustainability											
Emergency Management											
Delegates presentations											
Course Overall											

Note: Scores of 1 – 5 are from least satisfactory to most satisfactory

Figure 14.1 A typical training session feedback sheet.

Although the quantitative scores are very useful for senior managers to monitor how the training is going, the most important feedback is the written comments, particularly those constructive criticisms that can be used as a basis for continually improving the training content, materials and style.

Unfortunately, this cheap and cheerful method of measuring emotional reaction does not give much accurate information about how well people have understood the training and how rigorously they will go on to apply it.

If time is short, then an alternative method for getting "virtual feedback" is to set up a questionnaire using the Microsoft Forms website. This is free to use but to do this you must already have a Microsoft account. This can save time on the day of the training event, but does require more time later in chasing trainees to complete their online forms after the training is completed. The advantage of the paper-based feedback form is "you can't leave until you have handed in the completed feedback!"

The primary purpose of the Level 1 evaluation is to get feedback to allow future similar training events to be continuously improved. If the training is a one-off stand-alone event never to be repeated, then this type of evaluation may not be needed.

b. Level 2 training evaluation – Achievement of objectives

It goes without saying that most things in life have a purpose, and training is no different. The tutor must clearly identify the purpose of the training in advance by defining the objectives. These will be identified in the training session specification and can then be used to identify whether or not this particular training will meet the training needs of each potential trainee and also of the organisation. One of the common problems with training courses is that not enough thought has been put into the course objectives. Some years ago I was asked to carry out some training for a large multi-national company. This involved training in countries all around Europe. I had previously set up a "train the trainer" event but one manufacturing site did not send anyone to be trained as a trainer. The site manager of that facility contacted me and asked me to go to Switzerland to deliver the course to his personnel. I did not speak good French, and the local employees did not speak English, and so I pointed out that it was unlikely to be very effective. The response was that all the other sites had heard the same message directly from me and he wanted the same to happen at his site. It seemed to be immaterial to him that nothing that I said would be understood. In the end I had no choice but to deliver the training using an interpreter. That never works out well!

If you are unsure about the objectives, it can be sometimes useful to ask the individual trainees at the beginning, what their objectives are for being there – but be prepared for being told that their boss told them to come! I call these types the "Voluntolds!" (see Chapter 4)

If you ask people at the beginning to explain what their objectives of coming to the training are, then do not forget to go back to them at the end and ask them if their objectives had been met.

One of the objectives of the training will be to convey information to the trainee in a way that they both understand and remember it. Some organisations rely on some sort of test at the end of the training to check that key information has been remembered and understood. To do this in a meaningful way is quite a challenge. We must remember that workers who spend most of their working life doing practical tasks may not feel confident in writing long-winded answers. For this reason, such tests or quizzes tend to be designed for one-word answers or to use multiple choice answers. These can be quite effective – particularly in checking peoples' understanding of safety induction training, but in many cases that I have seen, the questions are so simplistic as to be almost valueless. If the objective of the training is to educate the trainees about the hazards of certain toxic substances, then a quiz or multiple-choice type of evaluation check may be appropriate, but if the aim is to train someone in how to use a flammable gas analyser, then a more practical test is likely to be required.

When preparing an evaluation or validation test always ensure that you check the difficulty of the test with a group of competent and experienced workers in the subject before you use it for the first time at the end of a training event.

If a validation test is used and there is a requirement for a "pass" threshold, then this should be made clear to the trainees from the outset.

c. Level 3 training evaluation – Behavioural change

The objective of any training event is not limited to providing the trainee with new knowledge and skills. What really matters is that trainees demonstrate their ability to apply those skills safely in the workplace and show that they have reached a satisfactory level of competence. This is known as an assessment of training transfer. This sounds very obvious, but actually is quite difficult to evaluate and measure. One way of achieving this is to observe and monitor how the trainee behaves after the training. Most organisations already do this when training apprentices, so why do we find it so difficult for other forms of training? This type of on-the-job evaluation is especially important if the training relates to how to carry out a specific task. It is somewhat less important if the training is of a general education nature, such as making staff aware of changes in recent health and safety legislation. The supervisor, first-line manager or a specially appointed "mentor" will have a crucial role in doing this. However, they need to have clear evaluation criteria against which they can assess the individual's performance and carry out a meaningful validation check. This takes time, because it will often need to be done on a one-to-one basis, and can only be carried out by someone who is already deemed to be competent in the subject. Remember that the objective of any training should not be just to complete the training – the over-riding objective is to get to the stage of confirming that people are competent. This will provide an indication of how well the training actually transfers into practice in the workplace.

Don't forget that there are certain types of health and safety training that fall into the "insurance" category, in that we hope that we will never have to use them. A classic example would be the training of first aid/CPR and the use of automatic external defibrillators. We train some people to do this, but we really hope that they will never need to use it. In this sort of training evaluation, it will be necessary to set up simulation exercises to assess how well the trainees have learned their craft – you can't wait to do a validation until an emergency arises!

d. Level 4 training evaluation – Organisational impact
What matters to senior managers is how a business or organisation performs. However altruistic a manager may be, ultimately, he or she wants to see a benefit arise as a result of spending time and money to improve. Training is a systematic way to improve the performance of employees. One of the managers' main objectives must be to see that the training that has been carried out, results in some sort of organisational benefit and ideally the manager likes to be able to look at some metric that assures him that progress is being made. There is often quite a long time-lag in realising these benefits and seeing results at the organisational level.

There are a number of measurements that can give an indication of whether the health and safety training is benefiting the organisation. Examples of such measures might include:

- Injury rates
- Near misses
- Procedural compliance
- Health and safety inspection actions
- Health and safety audit performance measures
- Employee climate surveys
- Number of breaches of health and safety regulation
- Number of safety improvement suggestions

All organisations periodically experience visits from the regulator. One of the predictable questions from the regulator is "how do you know that your employees are competent?" Records of health and safety training can also be very helpful to the organisation in these circumstances where they need to demonstrate that they have a structured approach to training and validation (see also Chapter 5). In the event of an injury at work it may be necessary to provide evidence that the people involved were adequately trained in order to avoid prosecution.

However, the best method of competence assurance is by the use of regular audits to ensure that workers are working safely and are aware that the management team is monitoring that performance.

Assessing Training Effectiveness Checklist

Assessment of training and competence needs to take into account the risk associated with the task. The higher the risk, the more detailed and thorough the validation checks need to be.

The Kirkpatrick assessment model:

1. Evaluate the quality and satisfaction with the training that has been carried out.
   (Typically done using a questionnaire at the end of the training. Very useful in assessing "enjoyment" and improving the training for future trainees.)
2. Measure achievement of training objectives. Usually done by tests or quizzes.
   (Ensure that objectives are clearly defined at the outset of training and that test questions are not superficial.)
3. Assessment of training transfer and application
   (A measure of competence. This needs to be done in the workplace by a supervisor or mentor.)
   Note that for training in rarely experienced events (e.g. Use of CPR) it will be necessary to include some simulation training.
4. Has the training had a beneficial effect on the organisation?
   (Measures can include such things as reduction in accidents or incidents)
   - Carry out regular audits

Chapter 15

Trainer/tutor selection – Select leaders who will inspire!

Once issues about trainee availability have been sorted, the next step is to ensure that there is someone available to carry out the training. Trainer, teacher, tutor, instructor, coach and mentor are all words that describe the role that we are looking for and are often used interchangeably for this purpose. Actually, they convey different meanings:

A teacher is someone who teaches or trains by giving information about a subject in order to expand the students' knowledge. Teaching is usually done in relatively large groups or classes in formal educational settings and follows a set curriculum. A teacher will normally have formal educational qualifications.

A tutor trains an individual or a small group outside the formal educational settings. The tutor works at the rate of the trainees and does not have an externally determined curriculum to follow. The tutor will often be a professional, but does not have to have formal educational qualifications.

An instructor, trainer and coach all deal with the practical application of knowledge in order to develop skills in the student that are of immediate practical use. Those skills involve acquiring and learning knowledge and applying that knowledge in a way that becomes second nature. They would normally be trained to a high level in a relatively narrow field of expertise. They would not normally have formal educational qualifications.

A mentor is an experienced and trusted advisor, who is willing to teach everything that they know to another less experienced individual. It is similar to parenting.

For the purposes of this book, we shall refer to the teaching being done by either trainers or tutors.

15.1 WHO SHOULD DO THE TRAINING?

So, who is likely to be person best suited to be a trainer/tutor? The tutor needs to be technically competent in the subject being trained. It is helpful if the trainees recognise the level of competence that the tutor has. One

DOI: 10.1201/9781003342779-16

problem that many young trainers/tutors have is that they might know the theory very well, but when trainees challenge them about practical difficulties, the tutor does not have enough practical experience to provide answers. This is OK if it is just an occasional lack of knowledge, but if it happens repeatedly, the tutor's credibility will be under threat. It is not unusual that some trainees can be very disruptive in these situations, by seeking to outsmart the tutor. Young tutors should always have the support of a more knowledgeable colleague until they have established their own credibility.

The nominated tutor should be someone who not only has the knowledge to carry out the training, but should also have the skills and patience to be able to conduct the training in a memorable and effective way. The tutor must have credibility with the trainees. Often this means that the trainer may come from the same peer group as the trainee.

Being a trainer should be regarded as an accolade. To be able to train others means that you have a level of skill or knowledge that many other people do not have. Conversely, it is also a requirement to have enhanced skills and knowledge if you are to be appointed as a tutor. As a result, I find that being a trainer is a wonderful way of developing your own skills. You cannot train someone else if you don't know your subject well enough yourself. In health and safety training it is very common that trainers may need to "brush up" their own knowledge level before the training because regulations and standards change so quickly.

The following diagram summarises the requirements that need to be fulfilled before appointing a trainer for a particular subject (Fig 15.1).

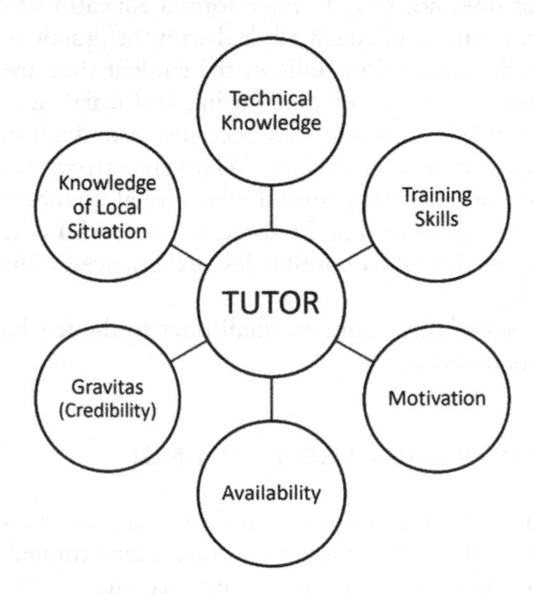

Figure 15.1 Criteria for tutor selection.

15.2 PRESENTER/TRAINER DEPUTIES

When preparing complex training courses, or if consultants are charging clients for carrying out training, it is wise to think about what should happen if a particular tutor is not available. This can happen for all sorts of reasons, most commonly because of illness or transport failures. If it happens during in-house training, the consequences may not be too severe, but if the trainees are coming from all around the world to the training event, they are rightly going to expect that there is a tutor available to train them. It is wise to ensure that, in the same way that you ensure that there are spare projectors available if the main projection system breaks down, that you have someone prepared, practised and available to cover the tutor role in the event that they are indisposed! This is mainly an issue for the longer type of training events which will probably have several presenters presenting different aspects of the training. In these circumstances I recommend ensuring that the presenters deputise for each other. This does not just mean telling someone that they are the deputy tutor for such and such session, but ensuring that they are completely familiar with all the training materials and what should happen and when.

Even if the training that you are planning is taking place "in house", always ask the question "What will we do if the tutor is not available or ill?" – even if the answer is as simple as that you will just postpone the training until another day.

Selecting the Tutor – Checklist

Ask whether the individual being considered as tutor for a training event:

1. Has the technical knowledge to tutor this subject?
2. Is suitably motivated?
3. Has suitable training skills and experience?
4. Has knowledge of local application? (This could be a problem when using consultant tutors)
5. Is available?
6. Has credibility with the trainees?

Chapter 16

Train the trainer – Showing others how to do it

In many larger organisations the sheer numbers of people requiring training can be off-putting and potentially very expensive. In these circumstances, if a standardised approach to a particular training subject is required, it is quite common to use the "Train the Trainer" approach. Train the Trainer is a distributed approach that creates a number of competent trainers and enables them to replicate a standardised training package.

This approach is very cost-effective and relatively fast. One of the problems with top-down training programmes is that they can be so cumbersome that they often run out of steam before all the required trainees have been covered. This can put the management team in a very difficult position if anything goes wrong that could have been prevented by the incomplete training programme. In effect the management have recognised a need for training that subsequently was not delivered. In the worst situations this might be deemed as legal negligence. The advantage of the "Train the Trainer" approach is that it establishes a network of competent trainers relatively quickly and cheaply, which means that completion of the training programme is much more likely.

The benefit of doing the training or instruction "in house" is that it is targeted and tailored towards the organisations needs and retains the training knowledge and skill within the organisation. It is often the case at the shop floor level that internal trainers are deemed to be more acceptable to the trainees than someone external.

The effectiveness of a "Train the Trainer" approach is dependent on two things:

a. The quality of the training materials
b. The selection of the local trainers

In many cases the local trainer will be the natural work group leader. However, these local trainers will need support, training and development themselves. Just e-mailing them a complex PowerPoint presentation for

DOI: 10.1201/9781003342779-17

delivery tomorrow will not guarantee a successful result. In selecting the local trainers, the minimum criteria should be to select individuals who have:

i. Good communications skills
ii. Sufficient experience and expertise to be able to answer questions on the subject
iii. Gravitas and robustness
iv. Time available to do the training

16.1 TRAIN THE TRAINER MATERIALS

The trainers who have been selected to disseminate the training will need training themselves in two areas. They will need to know the detail of the material that they have to deliver, but it is also likely that they will need some help in how to be an effective trainer.

The principle of "Train the Trainer" is that it uses basically the same materials and resources to deliver the same material concurrently in a multitude of different locations. Some of these may be overseas locations and therefore in different languages, but certainly in different national or regional cultures. I am a great believer in incorporating a wide range of diverse activities into training events to make them fun and to have high impact. The problem with this approach is that it can make them very complex to run. It is important to recognise that the more complex that you make the material that a delegated trainer has to use, the more difficult that it is for them to learn and the more things that there are to go wrong. As far as possible the materials to be used in "Train the Trainer" events should be kept as simple as possible. The training materials should be developed to be as user-friendly as possible to both the trainee and the delegated trainer. If, for example, in a PowerPoint presentation a question is posed to the audience, then instead of relying on the delegated trainer to know the answer, the answer should be displayed at the next click of the mouse. This way the trainer is able to give the same response as all other delegated presenters using the same training materials. Likewise, if the presentation involves hyperlinks or video clips, these should be set up to be initiated by the next mouse click and not require the trainer to have to remember to scan the screen for a particular hyperlink text or button. In other words, the person developing the training material should try and de-skill the delivery as much as possible to make it easy for the delegated trainers to use.

It is essential that the training developer also provides suitable tutor notes to allow the delegated trained to deliver the material. Some people prefer to have verbatim notes, whereas others like to just have key points identified and be left to choose their own words. Whichever route is taken, the training developer should prepare a "Trainers' manual" for each delegated trainer. My preference is that such a manual is prepared in hard copy format and is in a loose-leaf binder. The manual will contain the tutor notes. Tutor notes should contain a page for each slide in the presentation, which displays a miniature copy of the slide being displayed together with the tutor notes and any guidance that may be necessary for making discussion and interaction relevant. There should also be a clear description of the objective and arrangements for any exercise or activity carried out. The tutor notes can be easily generated directly from the PowerPoint slides by going to the "View" tab and then selecting the icon for "Notes Page". This will provide a miniaturised portrait layout copy of the presentation slide in the top half of the page and a text box in the bottom half to add the required notes.

The "Trainers' manual" should contain a master copy of the course handouts so that the delegated trainer can readily copy the materials that he or she will issue to their trainees. These handouts should include both the materials that will be handed to each trainee at the beginning of the training event, but also any other materials that are handed out during the training such as activity briefs and answer sheets. Where appropriate, the trainer's manual should also contain other local details such as any training record proformas and feedback forms which might need to be completed by trainees and any special instructions that are relevant such as fire evacuation arrangements. In some cases, it may also be appropriate to provide detailed question and answer information to assist the delegated trainers in responding to anticipated or unusual questions.

The trainer's manual will also include copies of any trainer's skill training information that has been provided during the "Train the Trainer" training event.

16.2 TRAINER SKILL TRAINING

It is assumed that the delegated trainers have been selected because they already have some appropriate knowledge of the area and topic concerned and that they have prior experience of training groups in their workplace. The skill training is therefore going to be related to understanding the purpose of this specific training and how it should be put over in order to achieve the training objectives.

The delegated trainers' skill training should clarify:

- Why the training is being done (objectives)
- Who is the target audience?
- How the training is to be delivered

- What supporting information is available to the delegated trainers
- What freedom do the delegated trainers have to change things
- The purpose and details of any practical work or exercises
- The timing and duration of the training
- What equipment/facilities are required

Presentational technique training should cover:

1. Preparation and hygiene requirements
   - Room layout
   - Equipment (is it all there and does it work?)
   - Course materials
   - Interruptions (are interruptions permitted to convey urgent messages)
2. How to involve the people in discussions
   - Eye contact
   - Pose questions
   - Use of flip charts
   - Encourage feedback from all
   - Check their understanding
3. Presentation style
   - Be enthusiastic
   - Vary the pace
   - Use personal examples and experiences to emphasise important points
   - Be prepared for the obvious questions
   - The importance of timekeeping

Any presentational material or guidance notes used during the delegated trainer's technique training should be included in the Trainers' Manual.

16.3 DELEGATED TRAINER'S PRACTICE

The "Train the Trainer" training will inevitably involve an explanatory run-through of the training material. Once the delegated trainers are all familiar with the material, each trainer should be validated by asking them to carry out a demonstration of how they would carry out the training. For large-scale training events using the "Train the Trainer" rollout process, it is advisable that the training developer monitors/audits the quality of the delegated trainer's performance from time to time.

"Train the Trainer" Checklist

1. Selection of people as "Trainers" need the following skills
   a. Good communications skills
   b. Sufficient experience and expertise to be able to answer questions on the subject
   c. Gravitas and robustness
   d. Time available to do the training
2. Trainer's Manual should be in a loose-leaf ring binder and contain:
   a. Full set of handouts
   b. Set of Tutor Notes for each slide
   c. Training record and feedback forms
   d. Any special instructions (fire evacuation arrangements, etc.)
3. The delegated trainers' skill training should clarify:
   a. Why the training is being done? (objectives)
   b. Who is the target audience?
   c. How the training is to be delivered?
   d. What supporting information is available to the delegated trainers?
   e. What freedom do the delegated trainers have to change things?
   f. The purpose and details of any practical work or exercises?
   g. The timing and duration of the training?
   h. What equipment/facilities are required?
4. Preparation and hygiene requirements
   a. Recommended room layout
   b. Equipment (is it all there and does it work?)
   c. Course materials
   d. Interruptions
5. How to involve the people in discussions
   a. Eye contact
   b. Pose questions
   c. Use of flip charts
   d. Encourage feedback from all
   e. Check their understanding
6. Presentation style
   a. Be enthusiastic
   b. Vary the pace
   c. Use personal examples and experiences to emphasise important points
   d. Be prepared for the obvious questions

Part II

Training resources section

Successful training is nearly always specific to the local needs. Even if it is a commercially available purchased package, it should be interpreted to make it relevant to the local situation and trainees needs. For that reason, it is impossible to recommend specific approaches to training without knowing the specific needs of the trainee and the facility.

The following resources are not intended to be a series of definitive, pre-prepared ideas that can just be cut and pasted into your training package. If they do suit your needs, you are very welcome to copy and use them, but they are really intended to be thought-provoking and demonstrate the sort of ideas that you might be able to adapt and develop for your own purposes. The important thing is that in order to have a high level of impact with your trainees, you need to think "out of the box" so that you can deliver activities that are relevant, unusual and that will be memorable.

Topics in the Resources Section

Page No.	*Ref No.*	*Training Resource*
	A	Jigsaws
	B	Photohazard spotting
	C	Camera Hunts
	D	Bespoke Videos and DVDs • Communications • Behavioural safety
	E	Flash Cards
	F	Gizmos • Consequences Game
	G	Interactive Exercises • Car Crash Exercise
	H	Games • Solway Risk Game • Happy Human Factors • Hierarchy of Controls Magnetic Darts

(Continued)

DOI: 10.1201/9781003342779-18

Page No.	*Ref No.*	*Training Resource*
	I	Role Play • Excalibur Supplies Ltd Case Study
	J	Event Case Studies • Lochside Engineering Ltd
	K	Emergency Simulation
	L	Competitions – Trainees' Presentations
	M	Quizzes
	N	Puzzles
	O	Mock Scenarios • Safe Systems of Work/Energy Isolation Plans • Confined Space Entry • Working at Heights
	P	PPE Exercise
	Q	Noise Simulation Exercise
	R	Manual Handling
	S	Crosswords and Piecewords
	T	Personal Commitment Statement
	U	Spot the Hazards
	V	Coincidence or not?
	W	Unfamiliar Task Assessment slide rule
	X	Microbooks
	Y	Circle the hazards

Where to find help when designing practical training activities

Training Topic	*Part I Page Reference*	*Part II Resource Reference*
Accident Investigation		A, J, K, G
Auditing and Inspections		C, D
Behavioural Safety		D, I, O
Cause and Effect		F, H(a)
Campaign Slogan Introduction		A, S, T, X
Communication		N
Confined Space Entry		O
Construction		I
Control of Substance Hazards		L
Emergency Management		J
Hazard Identification		B, M, O, U, Y
Hierarchy of Controls		H(c)
Hotel Safety		A, F2
Housekeeping		C, M

(Continued)

Training Topic	*Part I Page Reference*	*Part II Resource Reference*
Human Factors		D, I
Ice Breakers		A, H(a)
Laboratory Safety		B
Machinery Guarding		B
Management of Change		J, K
Manual Handling/Lifting		R
Noise		Q
Occupational Health		S, T, Q, R
Office Workstation Assessment		B, I
Personal Protective Equipment		P
Process Safety		B, C, J, Y
Risk Assessment		A, E, F, H(a), L, P, R, W
Safe Systems of Work		A, B, C, E, F, H(a), J, K, L, O, V, X
Safety Awareness		B, C, D, H(a), T, U, V
Safety Communication		D, N, S
Safety Sign Recognition		B, M
Scaffolding Faults		O
Sharing Safety Experiences with Others		L
Validation		M
Visitor Induction		F, I, T

Section A

Jigsaws

A1. Training applications:
(System or process flow diagrams/Risk Matrices/New campaign slogans)

A fun and competitive activity that can be used to introduce a complex idea before explaining it in detail to trainees. It is useful when introducing block or flow diagrams or introducing memory joggers or acronyms/mnemonics. Jigsaws can also be used as "ice breakers" in circumstances where the trainees do not know one another.

A2. Activity

a. Location – Separate "break-out" rooms or within the main training room.
b. Teams size <5 persons.
c. Preparation – Always check that all jigsaw pieces are there before training session commences.
d. Time – Typically a 80-piece jigsaw will take a team of three or four 20–25 mins.
   i. Set a time limit and award an incentive for the first to complete (e.g. mini chocolate bars).

DOI: 10.1201/9781003342779-19

ii. The time for the activity can be reduced by having all the pieces laid out in advance with picture side up. (This is only possible if jigsaw activity is in a separate break-out area.)

e. Make sure that once completed that the trainees actually read/ look at what the jigsaw says.

f. Caution – Once one team have completed their jigsaw, there is a natural tendency for the others to want to finish theirs. This can hinder the next part of the training if some trainees' attention is distracted.

Solution – Tutor helps lagging teams to complete their jigsaw.

g. Once the jigsaws are completed, it should lead onto a detailed explanation by the tutor.

A3. When producing jigsaws ensure that:

a. Jigsaw has no more than 80–100 pieces.

b. Ensure that it is practical (i.e. use a photograph or random pattern the as the background. Large areas of plain colour will render the jigsaw impossible to do.)

c. Put text in capitals and large font in text boxes with different coloured borders.

d. All lines should be minimum of 2 pt weight (thickness).

e. Files sent for printing into jigsaws must be in pdf format and not PowerPoint.

f. There are a multitude of bespoke jigsaw printers available on the internet at low cost.

A4. Example of Jigsaw

Assessing fire risk when visiting hotels:

Section B

Photohazard spotting

B1. Training applications:
(Hazard training/Good and bad practice identification/Workstation safety/Manual handling/Process safety hazards/Machinery guarding/Housekeeping)

An activity which tests trainees knowledge of a particular aspect of health and safety through the use of photographs. The photohazard spotting exercise takes place in and around the training room and involves the trainees moving around in pairs and matching wall-mounted photographs with the descriptions on a question sheet. The exercise is time-limited and a reward (mini chocolate bar) is given to the first pair to record all the correct answers. The clearly numbered photos are pre-prepared and located on the walls of the training

DOI: 10.1201/9781003342779-20

rooms. The trainees first have to find the photos and then match the photo to the description on the answer sheet.

Photospotting can be used for training in any subject where the learning/good practice/bad practice can be physically observed in a photograph. It has been used for such subjects as:

- Chemical hazards
- Housekeeping standards
- Visual display unit assessment
- Manual handling techniques

B2. Activity

i. Location

Best run in and around the main training room. To provide some interest and variety some photographs can be placed in adjacent rooms or corridors as this tends to get other people interested in what is going on!

ii. Teams

Best grouped in pairs to allow the trainees to discuss possible answers.

iii. Preparation

a. Take and select good quality photographs which will demonstrate the point you want to make. Select approx 15–20 to use in the game and print them with clear consecutive numbers at full page size (A4 or foolscap). If the photos are to be used repeatedly then it is a good idea to laminate them.

Example of Photohazard (Feet not touching ground)

Workstation Hazard Spotting

Spot the hazards on the photographs around the room and record the relevant photograph number on the table below:

Number	Hazard
	Poor posture
	Screen glare
	Arm rest set too high
	Restricted leg movement below desk
	Document rest incorrectly positioned
	Chair seat height incorrectly set
	Screen set too high
	Screen too dark
	Insufficient knee room
	Weights on back of chair
	Screen out of focus
	Screen set too low
	Mouse incorrectly positioned
	Inadequate wrist resting space
	Chair back incorrectly adjusted
	Body twisted

Question Sheet

b. Prepare a handout for the trainees which is a list of statements that describe each photograph and provide a column for the trainee to write their answer number in. The questions/statements should be in a random order.
c. Take a copy of the trainee's questionnaire and record the correct answers. Retain this for use of the tutor.
d. Stick the photos in a random order around the walls of the training room. It makes it more fun if some photos are outside the training room – in the corridor or break-out rooms, so that the trainees have to move about and look for them.
e. Caution – If the training is taking place in a quality conference suite or hotel, check in advance with the events manager that it is OK to stick the photos on the walls. If it is not OK then place the photos on horizontal surfaces or tape them to glass surfaces where they can do no damage.

iv. Running the exercise

Explain that there are a series of numbered photograph on the walls that relate to the topic being discussed. Explain approximately where they will be found (if more than one room). The aim of the exercise is to work in pairs to decide which photograph best answers the description on the question sheet. The first pair to get and record the numbers of all the correct answers are the winners of the chocolate bars.

The feedback to this exercise can be one of two options:

a. If time is limited go quickly down the question sheet asking the trainees to call out their answers – correcting them where necessary.
b. To get more detailed understanding, include copies of the photohazards in the next set of slides and discuss with the group what they thought and why. This approach tests the trainees understanding and gains more involvement but has the drawback of taking considerably longer.

v. Timing

The exercise takes about 15 minutes to complete. Feedback (a) will take about an additional 3 minutes, whereas feedback (b) can take up to 25 minutes.

Section C

Camera hunts

C1. Training applications:
(Hazard identification and awareness/Housekeeping standards/Machine guarding). It is particularly useful in situations where workers have been in the same environment for long enough that they either do not notice small daily changes or alternatively accept unsafe conditions because they have not caused a problem for some time. This is known as "visual bias" or "inattentional blindness" and is the cause of many incidents.

Camera hunts are a useful way to get trainees actively involved in hazard identification in their own workplaces. They are particularly valuable as a part of hazard awareness training or housekeeping improvement. Most people become blind to hazards in their own area if they are seen to be tolerated. This activity is very effective at raising awareness of a wide range of day-to-day hazards.

DOI: 10.1201/9781003342779-21

C2. Activity

i. Locations

This exercise is best done initially in the work location. Check in advance that there are no commercial sensitivities about taking photographs in the workplace or that the area to be used is not Baseefa certified or a designated Atex flammable area. Check also if there are any special requirements for permits to work. The teams will normally return to a central location (training or meeting room) to complete the feedback.

ii. Teams

Best done in groups of three, at least one of which should have knowledge of the location where the exercise is being done.

iii. Preparation

The preparation that is required is to have a fully charged digital camera for each group of three trainees, together with an easily understood way of downloading the photographs.

Ideally there should also be access to print large hard copies at A3 or small poster size. A simple PowerPoint template should be prepared in advance into which a photograph and text can be easily cut and pasted to allow the creation of a wall poster highlighting significant unsafe conditions.

There are three distinct parts to the exercise that can be run concurrently or in three parts on different days or shifts to suit the working pattern:

1) The camera tour	Done in the workplace
2) Feedback	Done in a training location or meeting room
3) Poster preparation	Done in a training location or meeting room

i. Running the exercise – The camera tour

a. The exercise should form a part of a broader training event. Small groups of three are issued with a camera. Each group should have at least one person who is familiar with the work area being visited. All members of the group must be reminded to comply with all the local personal protective equipment requirements and safety procedures.

b. The objective is to find examples of situations that are unsafe. This will normally default to being examples of unsafe conditions. Taking photographs of unsafe acts is more difficult and can lead to aggressive responses. Before taking photographs of people doing unsafe acts, it is advised that thorough behavioural safety training is undertaken.

c. Time in the workplace should be limited to about an hour. You will be astonished about what this activity finds. It is not unusual for groups to return having taken a hundred photographs.

ii. Running the exercise – Camera hunt feedback
  a. There are two options for feedback.
    i. All groups return to the training location and are asked to identify the photograph that they consider to represent the most unsafe situation. This situation is then described to all other groups of trainees. This photograph is then converted into a hazard awareness poster for publication around the workplace. A work request should also be prepared to correct the unsafe condition. The timing for this feedback is approximately 5 minutes for each group.
    ii. In order to gain the maximum learning, the alternative option is to share all or most of the photographs taken, inviting each of the three members of the group to contribute and share learning and recommendations with all the other groups. This approach maximises the learning but can be extremely time consuming, often taking over an hour to complete all groups feedback.

iii. Running the exercise – Poster creation

It will be evident from the camera hunt and the feedback that this exercise generates a wealth of valuable information. The challenge is to ensure that the heightened hazard awareness that has been engendered among the trainees lasts beyond the end of the training session and has some impact on others in the workplace. This can be helped by the last part of the exercise, which is aimed at publicising the unsafe conditions around the workplace. The idea is to use a simple PowerPoint template to allow each group to produce a simple poster drawing everyone's attention to the most significant unsafe condition. These posters are very powerful as they are not produced by management, but by the workers themselves. The group can write their own warning message and because a large number of photographs will have been taken, it is an easy task to change the poster regularly and keep everyone involved and engaged.

Each poster should contain:

- The photograph of the unsafe condition
- A learning message related to the unsafe condition
- The names of the group of three who produced the poster

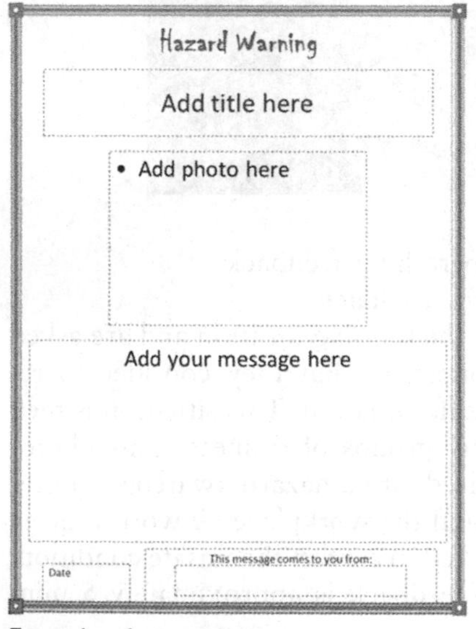
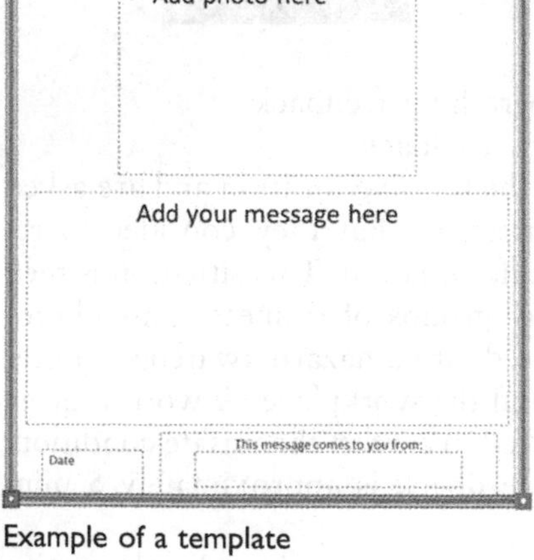

Example of a template

Hazard Warning

BEWARE OF MISSING FLOORPLATES

A missing floorplate beside the metal reject station could cause someone to fall and break an ankle. When it is necessary to remove floor plates, ensure that the area is barriered off and that access is prevented. Check your area to see if there are any missing floor gratings.

This message comes to you from:

"D" Shift Maintenance

Finished poster

iv. Timing

The workplace camera hunt should be limited to an hour. The feedback depending on which option is chosen can take from a few minutes to more than an hour. If a poster template is pre-prepared, the poster preparation should take about 45 minutes.

Bespoke videos and DVDs

D1 Training applications (General/Safety communication/Behavioural safety/Auditing observations/Remote learning)

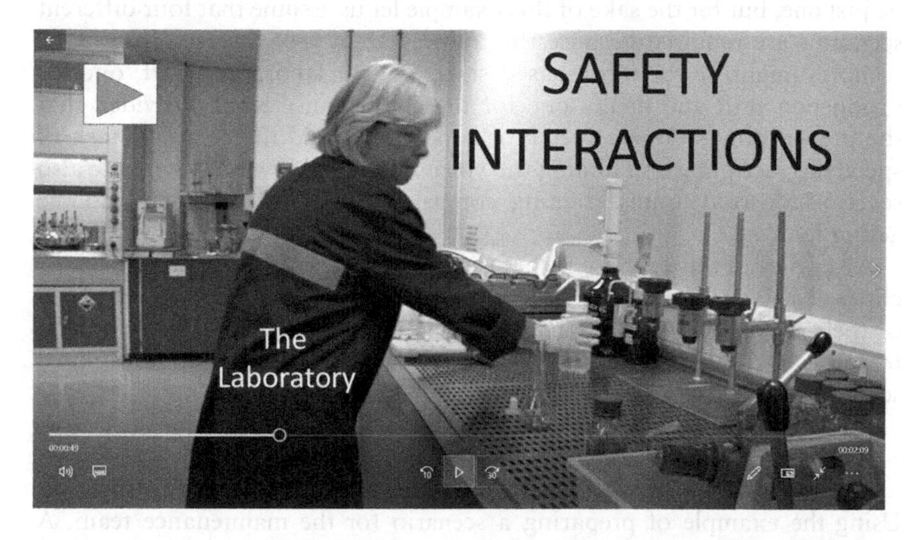

It is worth spending the time and expense of producing a bespoke video if there are to be a large number of repeats of a particular training event or if a scenario needs to be incorporated into online training. The use of bespoke videos is also appropriate if there are limitations on access to the workplace because of commercial confidentiality, potentially hazardous environments or because the training is being run in an off-site location such as a commercial training centre or hotel and there is a need to simulate the work environment during the training. It is also possible to use bespoke videos to portray either the safe way to do a task or to give examples of unsafe practice, but which is produced in a controlled environment that has been risk assessed. Tutors must always be aware of the risk that practical training itself can lead to an increased risk of harm because the activity is being done by trainees who at that stage are not fully competent.

DOI: 10.1201/9781003342779-22

The use of video clips is particularly useful in setting the scene for a practical case study or for a behavioural safety discussion or for safety communications training exercises. The video exercises would normally be carried out towards the end of the safety communication or behavioural safety training session.

D2 Activity: Video Example Option 1 – Scene setting video

The scene setting video is intended to provide the trainees with background information upon which they will then be asked to take some further action. Let us consider an example where the training is related to improving safety communication. As W.B. Yates reminded us "Think like a wise man but communicate in the language of the people".

Decide how many scene settings will be required. In most cases this will be just one, but for the sake of this example let us assume that four different scenarios are required because the trainees are a very diverse group. One scenario might be for office-based staff, one for laboratory staff, one for production staff and finally one for the maintenance team. Provided that the preparation is done correctly it is possible to do the complete task of specifying, scripting and filming the video in house. However, there are also several industrial drama training companies who will come and do the acting for you.

A key part of this exercise is that the actor or person featured in the video clip will also need to be available not only for the filming but also for each of the replays. It is important that during the training replays, the person featured is dressed in exactly the same way as they are in the filmed scenario.

Step 1 Prepare the scenario

Using the example of preparing a scenario for the maintenance team. A typical filming brief and risk assessment might be:

a. *Background*

Workshop is generally well maintained, but there are both safe and unsafe conditions.

Examples of unsafe conditions:

- *Trip hazard*
- *Aerosol left on top of oven*
- *Combustibles left near oven*
- *Sparks from grinding operation*
- *Pedestal drill with missing guard*

b. *The task*

Costume requirements

The technician (actor) will be one of a group of two or three people working in the workshop. For this reason, the actor will need to be dressed in the same overalls as the regular company technicians. He will be wearing light eye protection and safety boots/shoes.

Sound

The filming is unscripted and with background noise only.

Step 2 Control of risks during filming

It is essential that a suitable and sufficient risk assessment is carried out prior to filming. In this case the risk assessment identified the following precautions to be taken:

- *Oven will not be switched on and will be cold during filming.*
- *Grinding work is to be done by a qualified company technician.*
- *During grinding, the actor must have his back to the grinding machine.*
- *During drilling the actor will be seen to start the drill using the wall-mounted switch. The filming will then be cut, the drill will be stopped and filming will resume whilst the actor simulates the drilling task with the drill switched off. (The camera may need to be further away at this stage, so that it is not obvious that the drill spindle is not turning.)*
- *Filing will be simulated to avoid the tang doing any damage.*

Step 3 *Filming*

Sufficient film to be taken to produce between two and three minutes of final edited footage.

The filming will begin by a scan of the workshop to show both safe and unsafeconditions. The camera should avoid over-emphasis and not dwell for a long period on any specific condition.

The job to be done

The technician is working on a maintenance task. He has previously stripped down a pump and the work to be filmed involves placing some items in an oven in order to expand a sleeve to remove and replace a carbon bushing. While allowing the bushing to heat up he then goes to enlarge a hole in a flange section using the pedestal drill.

Details of job

The task involves manual handling of the pump onto the bench. The technician asks for help in lifting the item onto the bench (safe act). He then obtains a new gland bushing from the rack at the back of the workshop with bare hands (unsafe act) places it on the bench and puts on one glove before placing it in the oven (unsafe act). At this time there is another technician working nearby. He goes to use the grinder. The technician (actor) moves away from the area and then collects the part for drilling and goes to the pedestal drill. There is no flip-down chuck guard on the drill (unsafe condition), he has no goggles (unsafe act) and he is wearing gloves (unsafe act). The actor stops the drill and then after drilling he removes swarf and files the ragged edges (safe act) but has removed his gloves (unsafe act). The file tang is exposed and does not have a handle (unsafe act).

The technician leaves the area.

Props/preparation required

- *Gloves*
- *Aerosol container*
- *Pump*
- *Bushing*
- *Drill bit*
- *Removal of flip down guard on pedestal drill*
- *Old file with no handle*

An equivalent filming brief and risk assessment will be needed for each film clip (in this example there are four film clips used)

D3 Activity Exercise Option 1 – Running the activity

The exercise is intended to help trainees interact effectively with other workers. This is especially important when identifying and correcting unsafe behaviour.

- This exercise needs four mentors to lead each of four syndicate groups (see example of a "Mentor's brief" below).
- The exercise can demonstrate the communication skills of either one or two trainees /group/scenario . The example described here would allow 32 trainees to be tested if two persons interview at the same time, or 16 trainees if just one person demonstrates their interview skills.
- Break into four groups in four separate break-out rooms (the number of groups/rooms is determined by the number of scenarios filmed).
- Each group will move around the four syndicate rooms to witness each of the different workplace situations. For example:

- The "Office" situation
- The "Production Plant" situation
- The "Workshop" situation
- The "Laboratory" situation
- At the start of showing the video, the actor featured on that video is not in the room.
- After watching the video clip of the situation there will have been a number of both bad and good safety practices portrayed.
- As soon as the video is completed, the person portrayed on the video will walk into the room. Two of the trainees working together will hold a conversation with the person involved in the video about the good and not so good behaviours that they displayed in the video.
- The remainder of the trainee group will observe and provide feedback.
- After about 25 minutes, the group will move to the next syndicate room where a different pair will lead the discussion. Everyone will get an opportunity to be involved in the discussion!

D4 Example of the brief for one of the four session actors and tutors/mentors.

Safety Communications Exercise – Office scenario

(A similar brief would need to be provided for each of the four example exercises, so that the actors and tutors are properly prepared.)
Information on roles and locations

- *Actor – Catriona (playing Chloe)*
- *Mentor – Mark*
- *Room – Room 4*

Background to scenario

Chloe is a temporary agency secretary working in the company's purchasing office. She is a very experienced personal assistant, who feels that the work that she has been given to update the purchase order database is a little bit beneath her. She has worked at the company before, but this is the first time on this site. She was asked to stand-in for Susan who is on maternity leave. Apparently, the baby arrived early so the original plans for cover were overtaken by events! Chloe arrived on-site three days ago and is still getting to know her way around. She knows nothing about the business and her previous experience with the company was at the separate Head Office site. Because of Susan's sudden departure, she has had no real induction or training. Chloe is quite feisty and feels that she has been thrown in at the deep end. However, she is keen to show willing and wants to create a good impression.

Chloe has had no training in loading the photocopier, but even though there are lots of different makes and models she feels that it isn't rocket science. If she doesn't know how to do something, she tends to believe in trial and error. As a working Mum, Chloe is a great believer in safety at home, but she has never had any formal safety training in any of her previous employment. She considers it is much like being at home, but probably a bit more dangerous. One of her office colleagues told her about the fire evacuation test. The office alarm is tested every Wednesday at 11 o'clock. If the alarm goes off at any other time, then she has to go to assembly point in the corner of the car park, and someone will come and check that she is out of the building. She does not know what a toxic alarm is.

She has had no training in manual handling (i.e. lifting and moving heavy objects (like the boxes of copier paper), but carrying her little boy around does give her frequent back problems.

Issues:

1. *There is an open glass beaker of what appears to be a yellow chemical on the office window sill.*
2. *There is an open element electric fire on the floor behind Chloe's chair (she feels the cold and has brought this into work this morning from home).*
3. *Chloe's chair backrest and VDU screen are incorrectly set – this will aggravate her back problem.*
4. *The box of paper that she lifts into the filing cabinet weighs nearly 11 kgs. Lifting this from a high shelf could cause injury.*
5. *Chloe opens the photocopier door in an unsafe way.*
6. *Crouching down behind the door is asking for an injury.*
7. *Walking around with a cup of hot coffee could result in spillages of hot liquid and scalds.*

Purpose of the exercise

The purpose of the discussion is to allow the trainees to discuss both the safe and unsafe acts that Chloe is carrying out. It is not the objective to ensure that the delegates identify all the safe and unsafe acts, what matters is how the communication process goes. We are looking for the trainees to demonstrate:

a. *Explaining why they are there*
b. *Give credit for good practice*
c. *Confirm that it's OK to talk – if not arrange a convenient time*
d. *Allowing Chloe to do most of the talking*
e. *The use of open questions*
f. *Demonstrating that they listen and seek amplification*
g. *Ask Chloe about how he could reduce the risks associated with unsafe acts that he is demonstrating*
h. *Before leaving make sure that Chloe changes her behaviour in relation to the specific unsafe acts that has been identified*
i. *Thank Chloe for her time and her safe working*

Timings of the exercise

The time between syndicate sessions is 25 minutes. Groups will move from 1-2-3-4 Rooms

2 mins	*Select who will lead partake in the discussions*
4 mins	*DVD running time (max)*
7 mins	*Discussions with Chloe*
3 mins	*Group invited to feedback on what went well and what could be improved*
2 mins	*Actor feedback*
4 mins	*Mentor feedback on learning points (see bullet points above)*

D4 Activity: Video Exercise Option 2 – Scene setting video

This use of video material is particularly suited to behavioural safety training. It uses a scenario that the trainee will be familiar with. Typically, an everyday task is filmed in which workers are filmed carrying out their normal daily task. Unlike the previous approach described under "option 1" this approach needs far less preparation and less risk assessment. There should be no specific attempt to "stage" bad practice, but to merely record work as it happens and in effect replicate normal working.

Video recording is normally carried out by co-workers using non-specialist recording equipment. Typically, the finished film should be no more than 5 minutes long, although longer filming can be edited down to ensure that a variety of actions can be captured. Before filming, it is essential to explain to those workers who will be on the film, that the purpose of the film is to help train others, and there is no intention to try and catch people out. After the video has been recorded, it is important to show it to the people featured on the video so that they can have the opportunity to object if it makes them look foolish.

D5 Activity: Video Exercise Option 2 – Using the video

This type of video (descibed in para "D4") is normally used in behavioural safety training which is targeted at training people in how to interact with other workers in a non-confrontational way. By its very nature this should be done in small groups (usually more than six). This type of training always involves practical application of the training in a real-life environment and that means walking around in a workplace talking with people and discussing what they are doing and what safety precautions they are taking. The practical work can be very time consuming particularly if workers are working away from their base. Although it is essential to do some face-to-face discussion, the training time can be reduced by using the filmed scenarios. In this situation after the initial behavioural safety and discussion technique classroom training has been carried out, the video can be shown to set the scene.

The tutor then role plays the worker shown in the video and the trainees are asked to carry out a discussion with him or her to practise the discussion techniques that they have just learned in the classroom. It is a good idea for the tutor to make clear that he or she is now "role playing" by such actions as putting on a hi-vis jacket, hard hat or lab coat to suit the nature of the scenario. Each of the trainees is given the chance to carry out a safety discussion with the tutor. Ideally, a different video clip should be used for each trainee. The video clips should be meaningful to the trainees and portray a scenario that they can relate to. The advantage of this approach is that, if necessary, the same discussion can be repeated several times in order for the trainee to learn from the tutor's feedback and hence improve his or her skill before going out and practising on potentially sensitive real workmates.

Typical timing – approx 15–20 minutes per trainee/scenario

Using a 3–5 minute video clip, each usage per trainee will normally take:

Show video clip	3–5 minutes
Carry out mock behavioural safety discussion	10 minutes
Feedback from group	3 minutes
Feedback from mentor	3 minutes

D5. Activity: Video Exercise Option 3 – Other applications of self-produced videos

Homemade video recordings can also be used very effectively to train personnel in health and safety auditing training. This is particularly useful if there are problems in getting access to the workplace for some reason. This could because:

- The training is being done remotely from the main workplace (A training or conference centre).
- There are significant hazards for untrained people accessing that workplace.
- The activity is a transient one that is not often seen or repeated.
- The activity or condition is too remote or too high to get safe physical access.
- The activity requires specialised protective equipment that not everyone will have.
- The recording is to be used as part of an online or computerised training package.

The use of video recording in auditing observation training is particularly valuable, because by filming from a particular or unusual angle it is possible to safely simulate a hazard that does not exist (i.e. filming through the side of a scaffold access platform and giving the impression that there is no scaffold), or by using the zoom facility to view a high-level activity or condition from a safe distance.

The benefit of this type of "walk about" video recording is that it is possible to take large amounts of video film that then can be selectively edited and titled using simple editing software such as Windows Movie Maker. Final cuts of the video should be no more than 5 minutes long, but once produced, the video can, if necessary, be played over by the trainees more than once in order for them to gain the maximum learning.

Section E

Flash cards

E1 Training applications:

(Wide range of applications e.g. Risk Awareness/Hierarchy of Controls/ What is important in safety leadership/Noise hazard levels/Auditing questions/Hazard Study)

DOI: 10.1201/9781003342779-23

Flash cards are a very simple and quick technique aimed at ranking information in an order or sequence. They are particularly useful in promoting discussion or demonstrating the way in which different people have different perceptions of risk.

E2 Activity – using Flashcards

i. Location

Usually done within the main training room.

ii. Teams

It is important to get trainees involved in discussion about how to rank the cards. It is therefore recommended that this exercise is done in groups of three.

iii. Preparation

A series of strips of card are pre-prepared indicating the information that requires to be ranked. The example shown is where the flashcards are to be used to demonstrate the Hierarchy of Controls. The example here shows the hierarchy printed in its correct order on white card. The card is then carefully cut into the six separate pieces (these are known as "Flashcards"). Enough sets of flashcards need to be produced for the number of groups taking part in the exercise.

Eliminate the hazard altogether
Replace the hazard with something less dangerous
Use Engineering protection (like guards and ventilation systems)
Use trained and skilled workers
Have effective & up-to-date instructions
Use personal protective equipment
Workplace monitoring (eg noise or gas monitoring)

For the exercise to work, the number of flashcards for each exercise should be no less than six and no more than 15. The exercise works better with more cards, but this takes more time. The best application that I have used is in understanding risk perception. Here each flashcard represents one of 15 causes of death from smoking – nuclear accident. Up-to-date details will be available from your national statistics office.

As an example, the annual UK data for causes of death is as follows:

- Natural causes: Age 40–64 1:187
- Smoking 10 cigarettes/day 1:200
- Natural causes: Age 16–39 1:1400
- Accident at home 1:14000
- Accident on the road 1:17000
- Leukaemia 1:26000
- Homicide 1:80000
- Accident at work 1:83000
- Fire at home 1:135000
- Falling from a bed or chair 1:320000
- Sporting activity 1:400000
- Falling from a cliff 1:3400000
- Railway accident 1:3500000
- Struck by lightning 1:10000000
- Radiation from nuclear installation 1:20000000

It is interesting to note that people who are not used to working in higher risk environments will usually rate the risk of death due to a nuclear incident as much higher than the statistic shows.

E3 Running the exercise

The trainees are split into groups of three. This is most easily done in the main training room, but can also be done in break-out rooms. Each group is given a set of randomly sorted flashcards. Each group is asked to sort the cards into a column on the table that best represents the ideal sequence. For example, in the case where the cards are being used to raise awareness about risk perception, each group are told that each card represents a cause of death. They are asked to decide which is the most likely cause of death and which is the least likely and arrange the flashcards in a column on the tabletop showing their perception of decreasing risk.

What will become clear is that there is considerable debate within each group about which causes of death are more likely than others. When the groups have sorted their flashcards (usually 3–5 minutes) ask each group to call out which they consider to be the top two highest risk and also the lowest risk. It will become obvious that not all the groups will agree. The whole point of the exercise is to demonstrate that different people have different perceptions of risk. A dedicated smoker will tend to think that the risk of death caused by smoking is less than someone who has been bereaved because of their partner having lung cancer. Likewise, a skilled mountaineer will have a different perception of the risk of climbing to that of a novice.

The key message at the end of this exercise is that assessing risk is often a judgement made by fallible human beings and it is not absolute.

If the flashcards are used just to reorder or reinforce something like the Hierarchy of Controls, then the tutor can just check that everyone is in agreement and ends up with the correct order of cards.

Timing:

Ranking the flashcards usually takes the groups about 3–5 minutes. The tutor can use his or her judgement when to call time – this usually occurs when the tutor observes that most groupos have stopped shuffling cards around!

Section F

Gizmos

Gizmos are unusual high-impact devices for getting key messages across. They are often very memorable and a lot of fun. I usually would tend to make my own gizmo to suit a particular situation.

F1. Gizmo 1 – Spinning plates

F1.1. Application: The spinning plates activity is a fun way of demonstrating to managers in particular just how much effort has to go into managing safety. It can also be used as an "ice-breaker".

I am often asked by senior managers in particular "If there is one single thing that I could do to improve the safety in my area what would you recommend?" I usually respond "Unfortunately there is no silver bullet. Safety is like a garden, it needs tending in lots of different areas at the same time to succeed". In other words we need to keep a lot of "balls in the air at the same time". The spinning plates activity demonstrates to managers the challenge that they face.

F1.2. Preparation: Obtain enough spinning plates kits (plate and stick) for all attending. These spinning plates are readily available from the Amazon or E-Bay websites at a cost of around 2–3 dollars/kit. The tutor may need to demonstrate how to spin the plates so a little bit of practice is advised!

F1.3. Activity: Get everyone spinning their plates at the same time. After a short practice, offer a prize (mini chocolate bar) for the one who can keep their plate spinning longest. The biggest problem you will have is stopping people spinning their plates! Make sure that you relate the activity to the challenges of safety management – it needs a little attention all of the time to have a safe environment in the same way that the plate will stop spinning if you stop attending to it.

F2. Gizmo 2 – Consequences game

F2.1. Application: A fun way to demonstrate the consequences of taking risks.

DOI: 10.1201/9781003342779-24

F2.2. Location: Done within the plenary session in the training room. For this to work well it is best that the trainees are seated in a U-shaped layout rather than theatre style.

F2.3. Timing: The exercise takes about 5 minutes for a group of 15–18 trainees.

F2.4. Preparation: Need to prepare a "Consequences" game (risk allocator) which randomly allocates positive or negative consequences to the players' actions. The typical consequences used are:

- You got away with it this time	- No consequences
- You injured yourself	- Wear an arm sling
- You caused an injury to the person on your left	- They must wear an arm sling
- It's your lucky day and you finish the work early!	- You win a chocolate bar

Depending on the type of risk allocator you make, you may need to duplicate or add to the consequences

In addition to making a risk allocator, you will also need a supply of triangular bandages pre-knotted to form broad arm slings and a suitable quantity of rewards (mini chocolate bars).

Types of Consequence Game

i. Electronic game: My preferred device is a small electronic random generator. In my case it was originally bought as a "drinking game". I confess that I have never used it for that purpose, but by covering it with a simple printed overlay, I have converted it into the simple "Consequences Game" seen below. The player presses the button in the centre of the board and when the light stops it shows the consequences of their action.

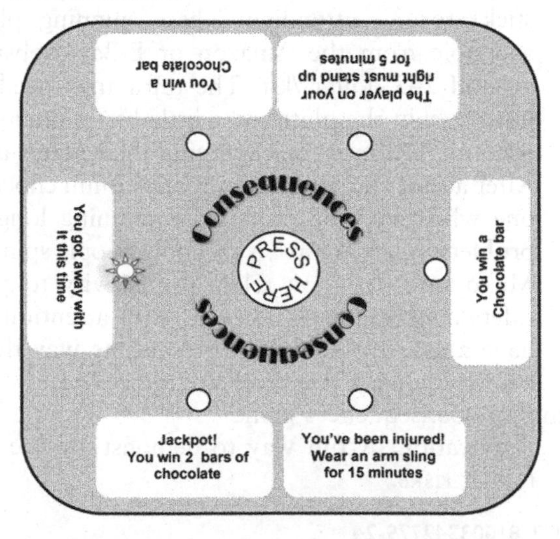

Although this particular "Drinking Game" is no longer available for purchase, many greetings cards contain similar random generators from which the electronics can be removed and converted into an equivalent "Consequence Game".

The advantage of this approach is that the consequences of taking the risk are truly random. However, the downside is that when used in a small group, it may not demonstrate all of the outcomes that the tutor wants.

ii. Playing Cards: I have also made a set of playing cards to demonstrate the consequences of taking risks. The same outcomes are used as for the electronic game. The cards are shuffled in the presence of the trainees and then fanned out face down and each trainee is invited to take one card blindly.

This approach is easier to prepare but not strictly random, but does demonstrate the fact that we take chances blindly. The advantage is that by limiting the number of cards to the number of players, the tutor can ensure that some people get the negative consequences and end up wearing arm slings!

iii. Hexagon spinner: The hexagon spinner is the simplest random generator. In this version of the Consequences Game, the consequences are made by spinning a hexagon dice.

To use the hexagon dice consequences game, replicate the hexagon shown below on white card and carefully push a pencil through the centre to make a simple spinner, ensure that the pencil is perpendicular to the hexagon when in use to get an unbiased result.

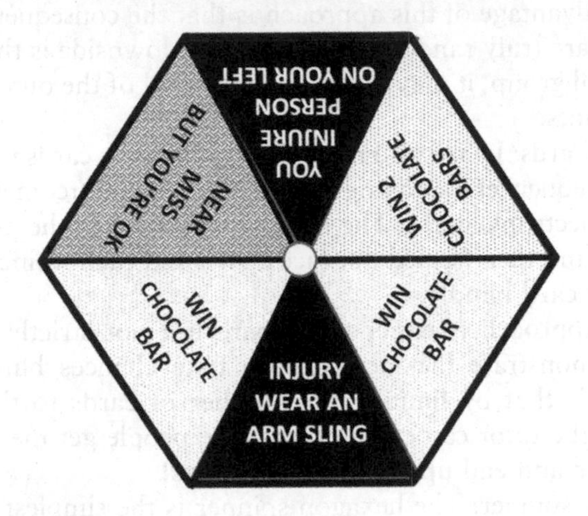

F2.5. Running the Consequences Game exercises.

Explain that you are going to invite people to take a chance. The consequences of taking that chance might be welcome (win chocolate) or unwelcome (you are injured and have to wear an arm sling for 15 minutes!). The exercise is done in plenary (i.e. without trainees moving from their seats). The tutor passes the consequences game risk generator to each person in turn and gives them their consequences (chocolate or arm sling) before moving on to the next. It is perfectly acceptable for a player to decline to take the chance – but this rarely happens!

At the end of the exercise which takes about 5 minutes with a group of 15 trainees, the tutor should summarise.

The game shows that:

- In reality people get away with taking risk more often than they are injured. That is why they perceive a benefit in risk-taking and continue to do it.
- Sometimes we take a risk and it affects someone else.
- Sometimes we take a risk and we don't even notice it because there are no perceived consequences at all.

Section G

Interactive exercises

G1. Applications:
Interactive exercises are one of the best ways of getting participation in training, particularly when it is done in groups (i.e. at training "courses"). The following example is used during training in accident and incident investigation. It is intended to demonstrate that a structured approach to identifying the underlying cause of an accident is much better than just brainstorming.

G2. Activity:

i. Location: This exercise is integrated into a training presentation. The activity is part run and then the tutor carries out additional training and returns to the exercise to provide the solution. Hence the exercise is run in plenary within the main training room.

ii. Timing: The exercise adds about 25 minutes to the presentation time.

iii. Preparation:
The exercise is best based upon a real incident, although fictional incidents can be used provided that they are realistic. It is essential that the incident is not known to the participants – if anyone has prior knowledge of the cause of the accident, the exercise becomes pointless as its intention is to demonstrate that jumping to a quick conclusion does not always identify the underlying cause. I find that road traffic accidents are the best examples to use, because people are familiar with these and it is unlikely that people will understand the detail of what happened. It is useful to have a schematic diagram of the location of the incident and to prepare all the background information so that the tutor can anticipate foreseeable questions. Questions that arise that the tutor had not anticipated are unlikely to affect the conclusion and so answers to those can be made up on the spot.

G3. Example of an Interactive scenario for investigating a road traffic accident

DOI: 10.1201/9781003342779-25

Incident Investigation *Road Traffic Crash Case Study*

Tutor's brief – This information is to allow the tutor to answer questions posed to him by the trainees. This information should not be volunteered or made available to the participants before the exercise.

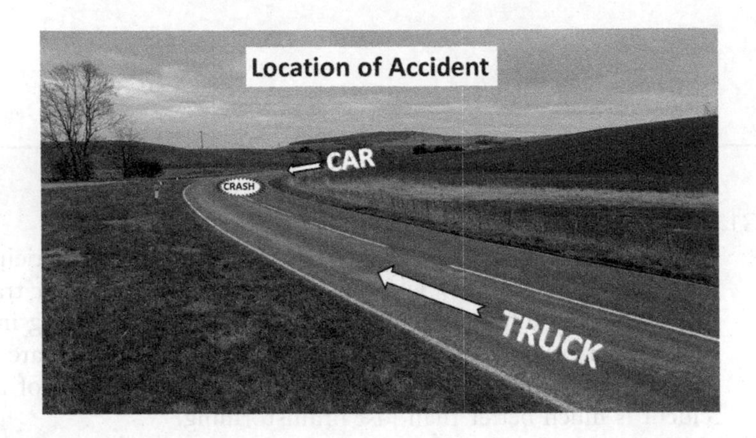

Background

Time:	*It was 11:00 in the morning.*
Collision:	*The accident happened in the United Kingdom where driving is on the left-hand side of the road. A car has collided head-on with a truck. All six people involved have died. At the point of impact, the truck is in its correct lane but the car has crossed the centre line of the road by approximately 3 feet (1 metre).*
Weather:	*A bright dry day but sun-blinding was not an issue.*
The car:	*The car was occupied by four persons. Mr Smith (aged 65) was driving and his wife Mrs Smith (aged 63) is in the front passenger seat. The two young occupants of the back seat are completely unknown to the Smiths. They are hitch hikers who have been picked by the car just 1 mile before the collision. Mr and Mrs Smith are very familiar with this road as they use it regularly. They had been driving this morning for about 40 minutes before the collision. Mr Smith has good health and recently had a medical check and was pronounced fully fit with no heart problems. An autopsy concluded that Mr Smith had not suffered from any sudden health problem.*
The truck:	*The truck is 10-tonne curtain-sider vehicle with a fixed chassis (i.e. not an artic). There are two occupants of the cab, the driver and his mate. They had been driving for 6 hours. In order for the driver not to breach his hours, the mate had just taken over driving. The mate had a heavy goods vehicle driving licence with no endorsements, but his licence did not cover this grade of truck.*

Car condition:	*The two-year-old car was a second-hand Austin and had been purchased by Mr Smith three months before the accident, but he had used it extensively since it had been purchased. Post-crash forensics showed that the brakes, steering and tyres were in good condition at the time of impact and there was no broken windscreen glass on the road before the crash site. There was no evidence of the Smiths being attacked by the hitch hikers (no weapons found).*
Truck:	*The truck was a five-year-old Volvo and had a full and up-to-date maintenance record. It carried a full load of drums of lubricant. No faults were found with the safety systems on that vehicle.*
The road:	*The road surface was clean, dry and generally in a safe condition. The collision occurred at the mid-point of a long radius bend. It appeared that the car had continued in a straight line as it entered the bend. There were wide grass verges on either side of the road – so sight lines should have been good. There was no evidence of animals (deer or dogs) that could have contributed. There is a manhole cover in the road about 150 yards from the point of collision on the car's side of the road. The manhole is slightly raised by about ½ inch (1 cm). This has occurred at some stage in the past when the manhole was lifted and then put back down on some trapped dirt. There are no signs of rubber skid marks on the road. The speed limit on this section of road is 60mph.*
Witnesses:	*There were no witnesses to the crash. A car following Mr Smith's car was travelling about ½ a mile behind and could see the Smith's car ahead. They were travelling at about 50mph and said that the Smith's car seemed to be travelling at about the same speed. They said that they did not see any brake lights go on prior to the collision.*

What really happened (this is revealed after the tutor has explained the technique of Root Cause Analysis):

The car was purchased second-hand three months earlier. When the previous owner was interviewed about the condition of the car, he mentioned that there was a problem with the interior bonnet (hood) release. The cable tended to seize and needed regular lubrication. It would appear that this information had not been passed onto the new owner. When the cable in the car was checked it was found to be seized, and the bonnet latch was in the open position. This car had a rear-hinged bonnet. The investigators surmised that the bonnet was being held closed by the secondary safety catch alone. When the car drove over the raised manhole cover, the jolt was not enough to affect the direction of the car, but was sufficient to release the safety catch, allowing the bonnet to blow open against the windscreen, obscuring the driver's view. This happened just as the car was entering the bend. Because the driver could not see the change in the road direction, the car carried on in a straight line crossing the centre of the road and directly into the path of the oncoming truck.

Running the exercise

Begin by displaying an image or photograph of the road where the crash took place. Indicate that this exercise is based on a real event but names and some details have been changed. Explain that the participants are the accident investigation team and that the tutor is their sources of information. Questions posed must be specific. For example, it is not sufficient to ask "Was there anything wrong with the car?" They need to ask things like "Was the braking system working correctly?"

Only volunteer the minimum amount of information to get the interaction going. Do not volunteer information about the fatalities – wait to be asked! One of the learning points is that participants get engrossed in the technicalities and forget that there might have been some casualties! They are usually stunned when quite late on someone asks if there were any casualties? The response of "Thank you for showing your caring side and showing concern for the people – actually 6 people died!" is very telling.

Be careful to ensure to stop the questioning before they identify the root cause. Questioning will get very lively, and I tend to bring it to an end by saying that "With all the substantial brain power that we have here today, we haven't actually arrived at the root cause. We have learned that brain storming or jumping to conclusions does not get us quickly to the correct root cause". At this stage the tutor should start to explain the incident investigation procedure using such processes as Root Cause Analysis, the Five-why process or whatever other system is used at the facility.

Once the system has been explained and the trainees have understood it, then return to the case study and help them go through the process systematically to end up with the root cause – which was the faulty design of the bonnet release mechanism.

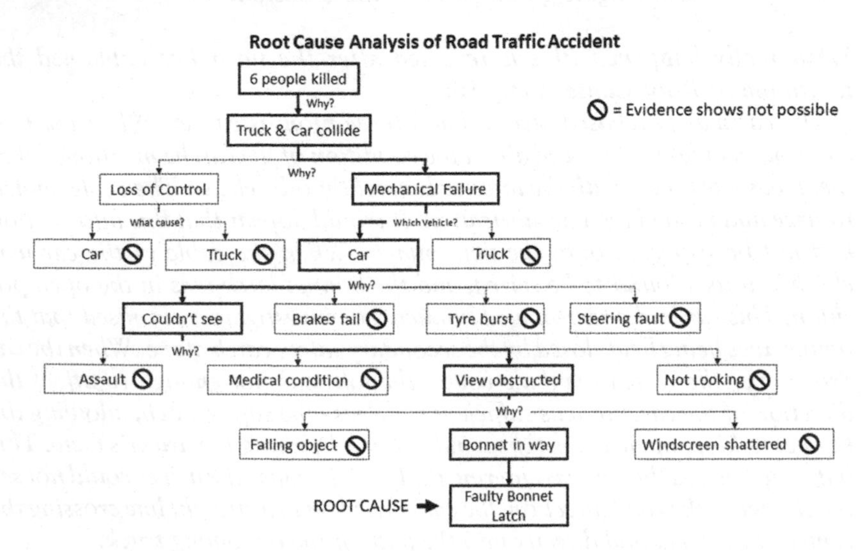

Section H

Games

H1. Applications:

Games can be a memorable and fun way of breaking the ice or reinforcing messages on such things as Safe Systems of Work. Can also be used as ice-breakers.

There are several commercially available health and safety games available by searching on the internet. Electronic games are not usually suitable for use by groups or during training course break-out sessions. Board games are most appropriate for this purpose. Care must be taken to check how long the game takes to run, otherwise it can adversely affect your session time management.

H2.1. The Solway Risk Game

This game is intended to reinforce the six steps of risk management, which are:

1. Look for the hazards
2. Identify who or what might be harmed
3. Carry out a risk assessment
4. Decide what extra controls are necessary
5. Record/communicate the findings
6. Review the risk assessment

DOI: 10.1201/9781003342779-26

H2.2. Location: Best run in a break-out room as things can get a bit raucous. There is also a need to get the game set out in advance.

H2.3. Teams: Suitable for 4–7 players

H2.4. Preparation: The game is a conventional board game, designed for 4–6 players per board. The principle of the game is that when landing on a green square the player is seen to have taken a safe action and this results in having to wear an item of Personal Protective Equipment. Landing on the orange square the player is deemed to have taken a risk and may be injured. This may have to result in wearing a bandage of some sort.

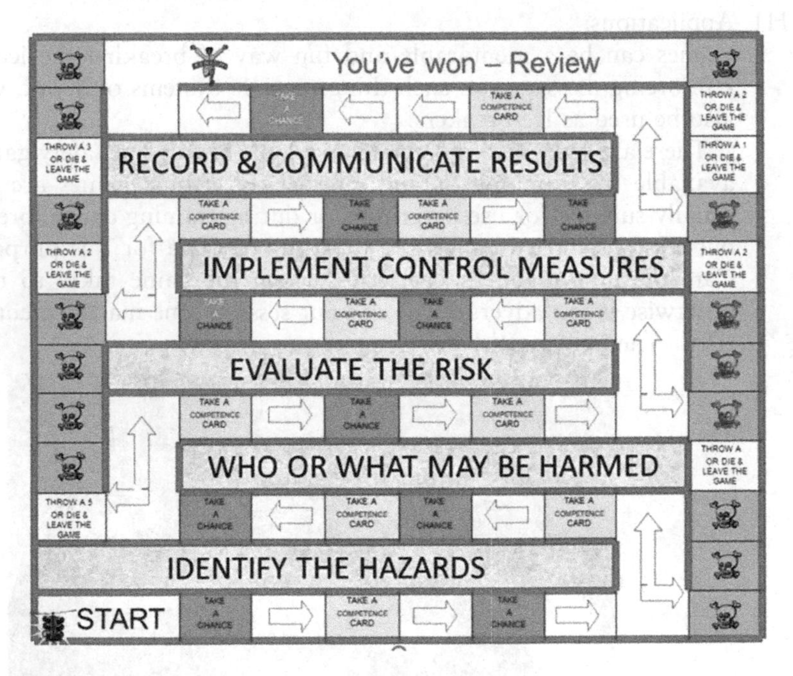

Each game requires the following equipment:
PPE

- Hard hats - 3
- Safety glasses (light eye protection) - 3
- Pair of riggers gloves - 3
- Disposable dust mask - 3
- Disposable earplugs - 3
- Eye patch - 3
- Broad arm sling - 3
- Elasticated head bandage - 3
- HiVis waistcoat - 1

Other items

- Dice and shaker container	- 1
- Players counters	- 6
- Quick explanation of rules card	- 3
- Set of competence cards	(see below)
- Set of chance cards	(see below)
- Chocolate bar for the winner!	

Each game requires two sets of cards to be prepared – one set of "Competence" cards and one set of "Take a Chance" cards. Much of the learning from the game arises from these cards, as they are read aloud when picked up. Ideally the two sets of cards should be printed on different coloured cardstock, with the words "Competence Card" or "Take a Chance" printed as appropriate on the reverse. For a typical game of 4–6 players, 15–20 cards for each set will suffice and the actions can be tailored to suit the message that the tutor wishes to convey. As an example, ideas for a set of "Competence Cards" and "Take a Chance Cards" are shown below.

Competence Cards

You failed to take account of people working nearby. You accidentally push over a pile of pallets which take the player on your left out of the game.	You are taking a chemical sample without any personal protection. Trapped pressure causes a splash and you complain of a burning in you eye. Wear an eye patch	There are no procedures for certain routine tasks. Operators are left to their own devices. Miss a turn.
You forget to record the results of your risk assessment. Go back one space	You are using a forklift truck as an access platform. The driver traps your arm against a steel girder. Wear an arm sling for the rest of the game	A new project is installed without a Hazard Study. Go back to the "Start"
One of the salesmen is sent to a Third World country without a health or travel risk assessment. Go back 4 places.	"Real men" don't worry about noise. You're losing your hearing but at least you can't hear you partner nagging you about unfinished chore! Wear ear plugs for the rest of game	You are working with a Permit to slip-plate a pump. You notice that the mechanical seal is leaking and decide to fix it. Miss a turn for failing to recognise a "change of intent"
Whilst removing rusty bolts, there is no eye protection available. It is only a quick job but the chisel splinters and you lose an eye. Wear an eye patch for the rest of game	You forget your hard hat and are stuck by a protruding scaffold pole. Wear a head bandage for the rest of the game	A new adhesive is used by the maintenance department, but without any assessment of whether it could release toxic vapour. Go back 1 space
Housekeeping is bad. You trip over an old pallet and break your arm. Wear an arm sling for the rest of the game.	You are working below a scaffold without a hard hat. You are hit on the head by a falling spanner. Wear a head bandage for the rest of the game.	You fail to carry out an adequate risk assessment of the risk of moving heavy blank flanges and injure your back. Get another player to move your piece for the next 2 turns and have a lie down

CHANCE CARDS

You decide that there is a risk of repetitive strain injury from moving your piece in this game. Your health is more important than winning. Volunteer to miss a go.	You have to unpack some newly machined spare parts. You decide that there may be sharp edges & decide to wear protective gloves for the rest of the game	Your are entering a construction area and following company rules you will wear a hard hat for the rest of the game..
You use a risk matrix to identify that although there is a significant chemical hazard, the risk of exposure is insignificant. The risk is tolerable, so you move forward 1 space	You were trained 10 years ago to change a grinding wheel, but you have not done one for 5 years. You recognise the need for re-training. Take another throw of the dice.	You use your experience to prevent a new employee injuring himself. He will be more careful next time. Give yourself a pat on the back & move forward 2 spaces
Your are working in an area of "low headroom". You do a risk assessment & decide that you will wear a hard hat for the rest of the game.	You are working adjacent to a public area. You identify that there is a risk to visitors & the public and decide to put up barriers. Go forward 2 places.	You need to go into the laboratory to retrieve your notebook. Follow procedures & wear eye protection for the rest of the game.
You are working in a construction area where mobile plant is operating. Assess the risk and select a suitable item of protective equipment to wear for the rest of the game.	You are working out of door and the air temperature is 5 degrees below freezing. You decide to wear gloves for the rest of the game.	You have to sweep the floor of the warehouse. You decide that there is likely to be a lot of dust. So you and the player on your right decide to wear dust masks.
You have just been validated as a qualified person for assessing risk. Have another throw of the dice	The maintenance team are exposed to oils & greases. You assess the risk of Dermatitis and insist that they use protective barrier cream. Move forward 1 place	You will be operating a tracked excavator for the remainder of the day and consider that the cab noise level is high. So you decide to wear hearing protection.

COMPETENCE CARDS

One of the delights of this game is that it can be a bit raucous and so it is advised that if there is more than one game board in use, that these games are run in separate break-out rooms. To minimise the amount of lost time, it is advised that the game board, PPE and all the other requirements are laid out on a large table in advance.

H2.5. Running the game

It is quickest to explain the rules of the Risk Game in the playing area. If there are more than one board in use, then it may be necessary for the tutor to have some assistance or have briefed a trusted trainee in advance.

Rules of the Risk Game

1. Follow the arrowed route from the start, via the six steps of risk assessment.
2. Throw the dice to start. Highest score starts then continue clockwise.
3. No extra throw of the dice if you throw a 6.
4. Do not cross thicklines or stop on a skull and crossbones.
5. The pink routes are chancer's alley. Use these at your peril – don't stop on a skull and crossbones or you are out of the game.
6. If you land on an orange "Chance" or green "Competence" square, take the appropriate card, read it out aloud and take the appropriate action. Replace the card at the bottom of the pack.
7. More than one counter can occupy the same square.
8. Winner is the first past the winning post.

This game is best run at the end of a presentational session such that it naturally leads into a coffee or lunch break. It is not ideal to run it in the middle of a presentation as trainees will tend to be a bit "excitable" when they return!

When clearing away after the event please remember that items like masks and ear plugs are single use only and should be disposed of. Elasticated head bandages should also be washed between uses.

H2.6. Timing: Allow 30 minutes for the game. It is best run just before a break so that if different teams take slightly longer than others it does not disrupt the training programme.

H2. Happy Human Factors

H2.1. Application: Used to supplement human factors training in health and safety and also in behavioural safety training. The game is an adaptation of the family card game Happy Families. The purpose of the game is to get the player to recognise the breadth of factors that can have an influence on health and safety performance. By playing the game participants learn:

- There are many more factors associated with "people" than you may imagine.
- You have no control over what hand you receive – no control of the consequences.
- You learn with experience/time and can influence the outcomes.
- It is more difficult if everyone is not open and honest!

This game is usually played part way through human factors or behavioural safety training session.

H2.2. Preparation: The game needs sets of bespoke playing cards to be produced in advance. The cards will have pre-prepared "families" of human factors that are influential in relation to good health and safety. For example, the sort of factors that can influence human behaviour at work might be:

- People
- The team
- The boss/leader
- The task
- Consequences
- Plant and equipment
- Skills
- Errors
- The organisation

The game takes each of these factors and has a family of five linked behaviours. So for example the Skills factors might be:

Training/Competence/Practice/Bad Habits/Supervision.

Each card will have the main Human Factor identified in red and the aim of the game is to collect all the five behaviour cards in the set of "Skills" behaviours.

The players learn to associate the five skill factors with the human behaviour of "Skill".

An example of the set of cards for the "Skill" behaviour is shown below:

Example of a full set of cards for "SKILL" factors

Ideally there should be about eight or nine "families" of behaviours for the game to work well. Ideas for a full set of cards are shown below:

Suggestions for a full set of Happy Human Factors Cards

SKILLS	*PEOPLE*	*TEAM*	*TASK*	*PLANT*	*ERRORS*	*ORGANISATION*	*BOSS*	*CONSEQUENCES*
Training	Behaviour	Goals	Resources	Safe Design	Skills	Culture	Dependency	Injury
Competence	Attitude	Pride	Tools	Ergonomics	Knowledge	Values	Coaching	Damage
Practice	Experience	Commitment	Conditions	Access	Misunderstood	Standards	Leadership	Death
Bad Habits	Capability	Contributor	Location	Shortcuts	Violations	Walk the Talk	Control	Near Miss
Supervision	Communication	Brother's Keeper	Preparation	Interfaces	Forgot	Auditing	Monitoring	Dismissal

These human factors can be changed to suit the local needs and messages. Each game can have up to six or seven players. The game can get lively and so it is best to have each game in a separate break-out area.

H2.3. Running the game

The game is best run sitting around a small round or square table. The pack of cards are shuffled and then dealt out around the players until there are no cards left. The aim is to collect full sets of behaviour cards. The winner is the player with the most completed sets when there are no cards left in any player's hands.

The game is intended to make you aware of a range of different Human Factors relating to EHS performance.

How to play.

The game is played like the traditional "Happy Families" card game, but in this case, the families are groups of human factors that can affect EHS performance. Each group contains five separate factors. For example, the group entitled "PEOPLE" factors contains the factors of:

Behaviour
Attitude
Experience
Capability
Communication.

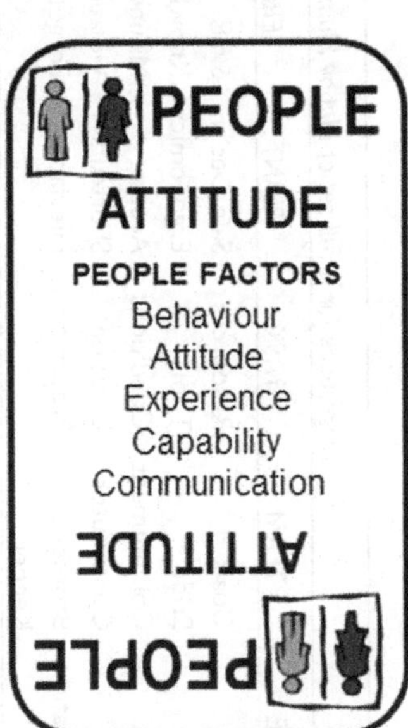

The group name is listed in large black print at the top and bottom of each card (e.g. "PEOPLE"). The contents of the group that players are trying to collect are listed in the centre of the card in smaller print. The individual factor for each card is printed in large letters below the group name (e.g. "ATTITUDE").

The aim of the game is to collect as many completed sets of Human factor "families" as possible.

All cards are dealt out to the players. Cards must be held so that the face of the cards is only visible to that player. The player on the left of the dealer starts by asking any other player for a card which will help towards completion of a set. The player (let's call him Pete) must ask for the group (family) name and the specific factor required. For example, "Joe, do you have the People – Attitude card". If Joe has the card he must give it up to Pete, and Pete may continue to ask for other cards either from Joe or from any other player. However, if Joe does not have the card, play moves to Joe and he may now ask any player for any card that he requires.

Once a player has all five cards in the set, the full set must be placed face up in front of him for all to see, but then play continues as normal. Play continues until there are no cards left. The winner is the player with the greatest number of complete sets of human factors.

The game takes about 20 minutes to complete. The fewer the number of players, the quicker it is.

- It sounds complicated, but four-year-old children play this!!!!!!!!!

H3. Hierarchy of Controls Magnetic Darts

H3.1. Application: Used for emphasising the sequence of the Hierarchy of Controls in the selection of the preferences for controlling risk. The Hierarchy of Controls requires that control measures are preferentially selected by following the hierarchy

1. Elimination
2. Substitution
3. Engineering Controls
4. Systems of Work
5. Behavioural Modification (Training and Auditing)
6. Personal Protective Equipment

The principle of the hierarchy is that when implementing control measures you should start at the top of the hierarchy by asking if it is reasonably practicable to eliminate the hazard. It is only if that is not practicable that consideration can move down the hierarchy to the next level and so on.

By changing the labels on the board to reasons why you are here, the game can also be used as an ice breaker.

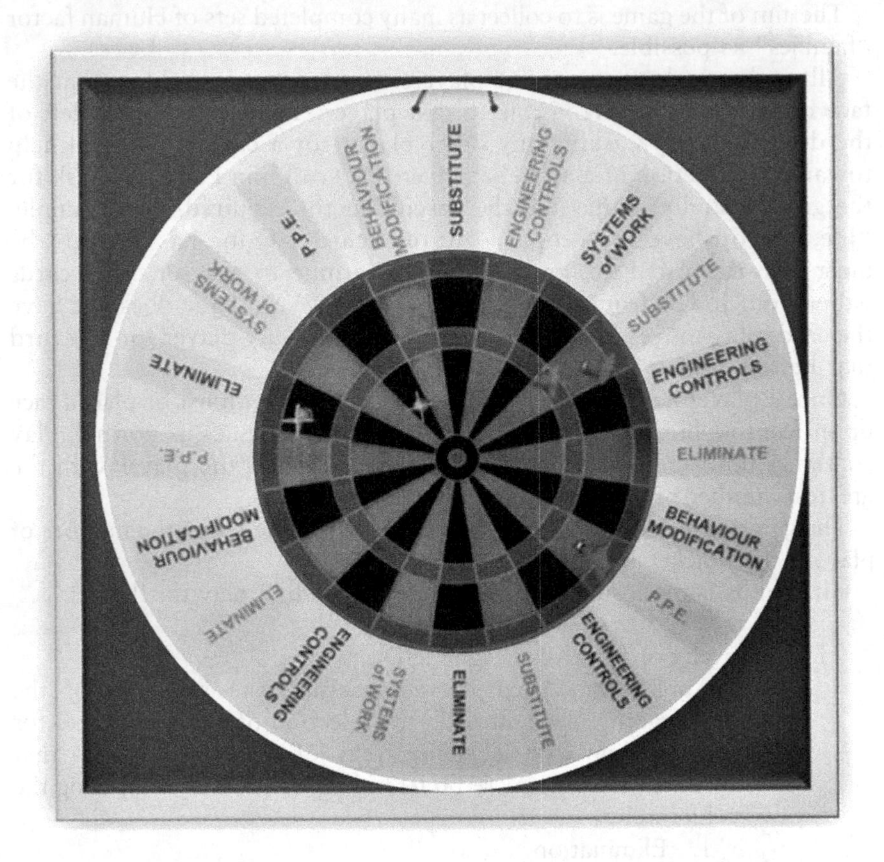

H3.2. Preparation: It is important to use a magnetic dart board (available from toy shops/websites). It would not be a good idea in health and safety training sessions to incur a puncture wound from a competition dart! The dart board can be mounted on a backing board and stickers highlighting the six levels of the hierarchy are stuck around the perimeter.

H3.3. Playing the game.

Divide the group into two or more teams. Teams play in rotation, one player at a time. Without using prompts, the player has to land the dart adjacent to the next heading in the hierarchy. The team that gets to PPE in the least number of throws is the winner. The game is quite quick and three teams will complete it in a total of 5–10 minutes.

H4. Converting proprietary games for health and safety training.

It is quite easy to create games that have health and safety learning. The photograph below shows a proprietary game called "The Magic Teacher". By simply changing the printed cover with a new overlay, the game can be converted to ask health and safety questions.

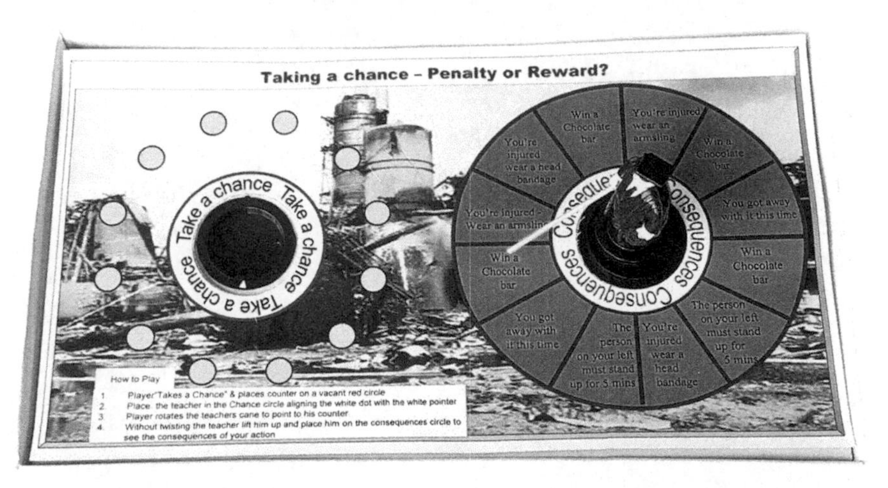

Section I

Role play

I1 Applications: Role play can be used for a wide range of applications where there is likely to be interaction between different individuals. These include Behaviour Safety discussions, Accident Investigation Interviews, Health and Safety Auditor discussions, Safety Communications.

I2 Preparation: Good role play requires preparation by the tutor and also for the main participants. A scenario needs to be prepared and this must be followed by a briefing note for the main players and an exercise brief for the other trainees. To avoid everyone having to read reams of information before the role play exercise can start, the more information that can be provided in visual form (drawings and photos) the better.

Example: Behavioural safety discussion role play

A. **Excalibur Supplies Ltd Case Study**

Summary (Available to all participants)

At 12:34 hrs on Thursday 19th March, Mr John Smith of Excalibur Supplies Ltd. was delivering a batch of new control valves to Colvend Chemicals International's (CCI) new Catalytic Converter project. While carrying a valve from his van to the project store, he was struck by a tracked excavator and trapped between a stack of pipes. The fire brigade took over an hour to jack up the excavator to free Mr Smith. As a result of his injuries, Mr Smith is now unable to work and is confined to a wheel chair.

Background – The Catalytic Converter (CC) Project

- *This is a construction site, where progress has reached the stage of groundworks nearing completion.*
- *None of the new plant roads are yet complete and the ground surface is churned up and muddy as a result of recent rain and mobile plant movements. Workers and mobile plant typically take the safest route that they feel to be appropriate.*

DOI: 10.1201/9781003342779-27

- *The project store has only one storeman. During his meal break from 1215 to 1245hrs, the project store gates are locked. CCI's order on Excalibur stated in "Condition of contract 27 (c)" that no deliveries will be accepted during the hours of 1215–1245. This order was processed by a new young sales clerk.*

What happened:

Mr Smith arrived at the CC Project site to find the project store gates locked. He entered the site by the next gate and parked beside the gate office. The security attendant was away on a "comfort break"! Although there was a sack barrow in the van, because the ground conditions were difficult and he would have to take a long detour, John decided to carry the first valve weighing 12Kg and take the most direct route to the project store. There were a set of open holes in the ground and his previous experience told him that these were a serious potential hazard and so he diverted around a pipe stack (3 tiers of 250 mm dia pipes x 10 m long) alongside a working tracked excavator. The gap between the excavator and the pipes was about 1.5 m. Just as he walked between the machine and the pipes, the excavator slewed and the rear counterbalance weight crushed Mr Smith against the pipes.

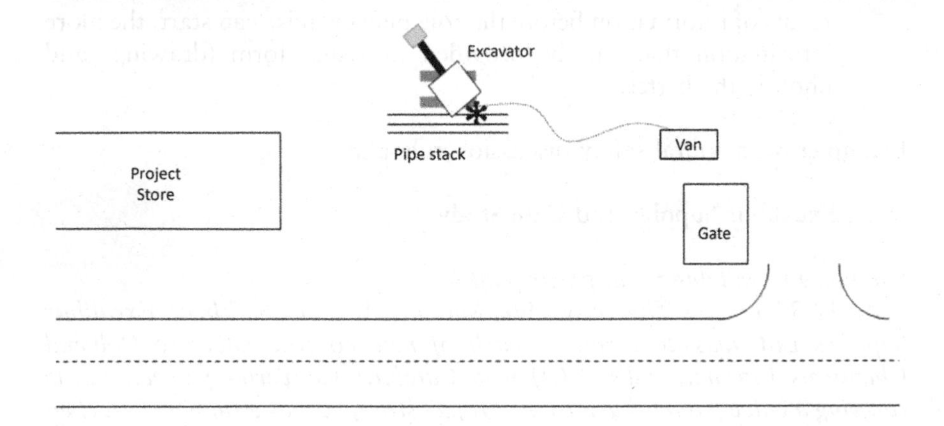

B. **Brief for Mr John Smith (Only issued to the individual role playing John Smith i.e. the interviewee)**

(Feel free to embellish the details to make it interesting, but don't change the basic facts)

Personal details:

- *22 years old, living with partner Sue. One 15-month-old baby boy (Clarence)*
- *Normally a final year Social Sciences student at the Local University*

Working for Excalibur Supplies Ltd.

- *Worked for Excalibur since last Monday (four days) during the University Easter vacation.*
- *Keen to make a good impression, as there is the possibility of more work during the long summer vacation.*

What happened on 19th March

- *I arrived at the CC project store at 12:28hrs and found the gates locked.*
- *I was running about 25 mins late because I got lost trying to find the CC project site. (First time I had made a delivery here).*
- *Parked by the security office – didn't check in as there was no one there.*
- *I had three valves to deliver and would normally use the sack barrow in the van, but the ground was too roughand muddy to use the barrow, and in any case I would have had to go a long way round.*
- *I was worried about what my boss would say as I was already late for my next delivery.*
- *I decided to carry one valve at a time directly to the store, as it was quite close by. There were no defined walkways, so I set off towards the store.*
- *I was aware from previous experience that excavations are one of the main hazards on construction sites. This is why I diverted around the holes on the CC project site.*
- *I went around beside the pipe stack to avoid falling into the excavations.*
- *The excavator's engine was running but the arm was not operating. There seemed to be plenty of space between the pipes and the excavator.*
- *The excavator's tracks were parallel to the pipe stack and to the way that I was walking and I thought that I would be all right even if the machine moved, because these tracked machines move quite slowly, and anyway we were facing in the same direction.*
- *I never thought that the top of the excavator would slew around.*
- *I assumed that the excavator driver had seen me.*

Experience and training

- *Worked as a labourer on construction sites during previous vacations.*
- *I received no site pass of safety induction to the site.*
- *I was not wearing any form of protective equipment, as Excalibur did not supply any. They did not expect me to actually walk around construction areas.*

C. **The syndicate task brief:**

Allocate roles of:

1. *Interviewers*
2. *Mr John Smith*
3. *Observer (to provide feedback)*

The interviewers' task:

Discuss for no more than 15 minutes with Mr Smith what happened and why did he do what he did?

Mr John Smith's task:

Using the brief provided, role-play the responses of Mr Smith during the interview. (Note: John Smith is the primary role play. He should have been given the Summary brief to study in advance.)

The observer's task: (Note that the Observer does not take part in the interview discussion, they observe what went rightand what didn't go so well in the interview and feedback at the end of the exercise.)

Identify what were the human factors that caused this injury to occur and how could behaviours have been changed to prevent it happening again?

I3 Location: Usually run in a break-out room

I4 Teams: The Excalibur Supplies Ltd Exercise uses teams of three or four

I5 Timing:

- 10 minutes for team to read brief and prepare
- 15 minutes for interview
- 10 minutes for observer's feedback

Section J

Event case studies (Lochside Engineering)

J1. Applications

Case studies can be used for multiple training purposes. The following case study is just one example of how it can be used for demonstrating failures in Safe Systems of Work, Confined Space Entry, Changes of intent, Control of modifications, Accident Investigation or Understanding Corporate Health and Safety Responsibilities.

J2. Activity

Case studies can be based on factual events or entirely fictitious. The simplest way of getting a realistic event with clear learning outcomes is to use a real event. The event should not be something for which the trainees already know the conclusions as this will mean that they can jump to the conclusion without the need for any detailed thought or analysis. If you wish to use an in-house example that the trainees are already familiar with, then it is advisable to change some of the details to suit the conclusion that you wish the trainees to reach. There are many major incident reports that are available for purchase at modest cost from the national safety regulators and these can be used as a basis for case studies. If you choose to write a fictitious case study, be careful to ensure that it is realistic.

J2.1. Training Location: Usually run in break-out rooms.

J2.2. Teams: The actual size of the teams will depend on the content and exercise brief. The team's size should be sufficient to allow for discussion and to allocate investigational responsibilities among the team members. In the case of the "Lochside Engineering" example shown below, the team size would be approximately six persons.

J2.3. Preparation

DOI: 10.1201/9781003342779-28

To develop a realistic case study takes a certain amount of preparation. In order to tell the story effectively you may need to provide some of the following information:

i. A task brief.
ii. Background to the incident
iii. Incident report
iv. Permits to work
v. Plan of workplace
vi. Engineering drawings/process flow diagrams
vii. Photographs
viii. Witness statements (written or audio recordings)
ix. Relevant procedures or instructions

One of the potential problems of providing sufficient detail for the case study is that in a thoroughly prepared study, there can be a lot of information. The more information that is provided, the longer it will take for the trainees to assimilate all the information. In these circumstances, the trainees should be given the information in advance of the training session such as pre-course reading, so that they come with a clear understanding of what happened.

The following case study (Lochside Engineering) shows the sort of preparation that can go into a case study exercise. If you wish to use this exercise you will need to copy the text in italics and also those shown as figures.

J3. Lochside Engineering

This case study is designed to be run in groups of 4–6 trainees working in a separate syndicate room. The time for the group working is approximately 1 hour with additional time allowed for feedback to the tutor or other groups. The event is fictitious and any resemblance between the characters mentioned and persons living or deceased is entirely accidental.

J3i. Possible exercise briefs

LOCHSIDE ENGINEERING LTD CASE STUDY

Group I Task

You are a team from the Health and Safety Regulators office who need to address the following questions:

1. *Were the injuries sustained in this incident foreseeable in carrying out work on a tank that had never stored flammable materials, and if so why?*
2. *What other injury risks had to be guarded against in order for this task to have been completed satisfactorily?*
3. *What would you have done to ensure that injuries did not happen on this job?*

Be prepared to present your responses to all three points on a flip chart to the other groups in no more than 5 minutes.

If there are more than one syndicate group working on the same case study, more learning can be achieved by giving the groups slightly different tasks. So, a second group might be asked:

Group 2 Task

You are the Board of Directors of the Lochside Engineering Company. You intend to find out:

1. *Who should have been in overall control of this job?*
2. *What was the responsibility of Lochside Engineering Company management?*
   a. *What was the responsibility of GRPM management?*
   b. *What was the responsibility of James Henderson?*
   c. *What was the responsibility of Dave Carter?*
   d. *What was the responsibility of Demon Hill?*
3. *Who should have carried out the risk assessment and what responsibility did the various people have regarding changed circumstances during the job?*

Be prepared to present your responses to all three points on a flip chart to the other groups in no more than 5 minutes.

J3ii. Background to the incident

LOCHSIDE ENGINEERING LTD

(For training purposes only)

Background

Lochside Engineering Ltd (LEL) are a medium-sized manufacturer of high-specification valves and fittings for the oil industry and steam and boiler plant applications. The company operates from its North

Eastern base at Colvend. The site, which covers 3.5 acres, employs 150 engineering staff, mainly working on days, but with a small shift team of two men/shift with a dual role of security and boiler plant operations.

The factory boiler is 35 years old and provides steam at 250 psi and 50 psi for a small turbo alternator, product testing and space heating.

The Incident

During the company holiday shutdown last August, opportunity was taken to carry out major maintenance and repairs to the factory steam systems and boiler. Part of the water treatment for the boiler feed water entails de-ionisation which requires regeneration with sulphuric acid and caustic. The sulphuric acid is stored at a 40% concentration in a glass fibre tank. Previous inspection of the acid tank had shown that it was in need of repair, and this repair was scheduled during the works shutdown this August.

The repair work was carried out by GRP Maintenance Ltd., a small local company with just six employees, but with previous experience of repairing this acid tank. The repair work started on Monday 13th Augustafter GRPM personnel had collected the necessary Permits to Work.

At 14.52 hrs on Monday 13th August, a fire occurred in the sulphuric acid storage tank and Mr Jimmy Henderson sustained serious burns which required hospital treatment, including extensive skin grafts and an absence from work for over five months.

Taped witness statements are available from the following:

- *Iain Wachem* — *Services Supervisor, Lochside Engineering Ltd*
- *Jimmy Henderson* — *Senior Technician, GRPM Ltd (the injured man)*
- *Dave Carter* — *Technician, GRPM Ltd*

The following personnel were also involved:
Lochside Engineering Ltd -

- *I Harding* — *Manufacturing and Operations Director*
- *J Watt* — *Works Engineer*
- *D Hill* — *Boiler Plant Operative*

GRP Maintenance Ltd -

C Swift — *Proprietor*

J3iii. Permit to work

	ENTRY CERTIFICATE	001300

NOT VALID WITHOUT A WORK CONTROL PERMIT

1
Plant: LOCHSIDE ENGINEERING LTD. Vessel/Confined Space: BOILER PLANT H_2SO_4 STORAGE TANK

Scope of work to be done: To carry out internal repairs as required.

2

Location of Tests	Oxygen %	Explosimeter LEL %						
ENTRY MANHOLE	21	N/A						

Analyst: Print Name: I. WACHEM/J.WATT Sign: I Wachem Time: 0800 HRS Date:

Note: To save yourself work when preparing documents to use repeatedly in training sessions avoid putting information on those documents that are date sensitive and that may need regular updating.

J3iv. Witness statements (to add interest these statements can be audio recorded and then played back during the syndicate exercise)

WITNESS 1 – IAIN WACHEM SERVICES SUPERVISOR, LOCHSIDE ENGINEERING LTD

My name is Wachem. I am the Services Supervisor at Lochside Engineering Ltd. I have worked at Lochside for 25 years and have been supervisor for the last 18 years. I am responsible for the operation and maintenance of the Boiler House, including the de-ionisation plant which includes the sulphuric acid storage tank. My boss Jamie Watt and I have been aware of the need to repair the acid tank for some time and Jamie had arranged for GRP Maintenance Ltd to do the repair. We use them because they have done good work for us in the past. We know their man Jimmy Henderson and think he is a good tradesman.

I arranged with the "C" shift boiler operator on night shift on Friday night to neutralise the remaining acid in the tank with caustic. We had calculated that there was less than half a metre cubed of acid left in the tank as we had been running the stocks down. After testing the neutralised acid we found the pH was 7.4 and so we dumped it to drain before flushing the tank several times with water.

On Sunday night the operators spaded off all the lines to the tank and removed the manhole cover. In preparation for the GRP men on Monday morning. I came in early at 6.30 on Monday to check that all the pipeline isolations were correct. Once I was satisfied, I called Jamie over to do the air test. I never do the test myself as I like to use someone else as a double check on the safety standards. After all, two heads are better than one.

The entry permit showed the air test results were 21% oxygen, which meant that the air was breathable.

Having completed the entry permit, I completed the Permit to Work. By this time Jimmy and Dave were on site and were pressurising me to let them get started. I told them to go and have their breakfast and that we would be authorising the work to start at about 10 o'clock.

I based the job descriptions on the method statement that had been provided by GRPM on the 15th of March. This made no mention of using solvent for degreasing. I had absolutely no idea that Jimmy had solvents inside the tank.

While the work was going on in the tank, I had asked Demon Hill, one of the boiler house operatives, to offload some sacks of spill-dri from a trailer at the back of the boiler house. I didn't think that this would have any effect on the tank repair work. At about one thirty Demon came to see me and said that he would have to stop the off-loading because the "GRP guys were being difficult and wanted him to stop". At that time, I knew nothing about the airline incident. I said to Demon that I would go and sort-out Jimmy and Dave.

I got delayed by a telephone call, and so went out to the back of the boiler house just before two o'clock. Just as I came into view of the acid tank, I heard a noise like the boiler firing up and saw a jet of flame shoot out of the acid tank manhole.

Dave Carter was trying to pull Jimmy out of the manhole. Jimmy's clothes were on fire. The flames in the tank only lasted for a few seconds. I called to Demon to get the first aider from the offices and to call an ambulance, whilst I helped Dave put out the flames on Jimmy's clothes. He looked pretty badly burned to me.

The first aider and ambulance arrived together about 5 minutes later. After the ambulance had gone, I called the fire brigade because I was worried that with the tank being fibre glass the whole thing might catch fire.

WITNESS 2 – JAMES HENDERSON SENIOR TECHNICIAN, GRPM LTD

My name is James Richard Henderson. I am 45 years old, and have been working for GRP Maintenance for over 10 years.

Most of the work that I do is involved in repairing tanks like the one at Lochside engineering: sometimes they are bigger and sometimes they are installed in the ground.

I am quite used to working inside old tanks and often have to use an airline mask because of the fumes and smells inside the tanks. While I have been working at GRPM I have never received any training about the safety hazards in confined spaces, although some years ago the boss brought in a consultant to talk to us about safety laws and stuff.

When I am doing repairs to the GRP linings in the tank, I know that the Attac resin won't bond to the tank if there is any grease about. I do the best repairs in the company – I'm always going back to replace the other lads' work, but my work will still be there when we're all pushing up the daisies! The secret is in the degreasing. I always use a solvent, I think it's called acetone. The cleaning is done before any matting or resin is put in place.

Last Monday Dave Carter and I arrived at 8 o'clock and went to see the Lochside Services Supervisor to get the OK to start work in the tank. Iain Wachem, the supervisor, told us that the man from the offices was just doing an air test, so we took our gear around to the back of the boiler house and started to get things ready. By about "breakfast time" – that's about quarter past nine, we had got our air masks, air compressor and equipment set up. The manhole door was off the tank, so we took a quick look in without going inside. It was very dark with no natural light. This is unusual as fibre glass tanks often let some light in. We didn't have a light, so Dave went to the electrician's workshop to see if we could borrow one.

After breakfast we got the permit from Iain to start work. We started the air compressor and then I put on my chemical suit, helmet, BA mask and gloves and climbed into the manhole. Rab stayed outside the tank and passed me the electric grinder and light.

It took about an hour and a half to grind down the damaged areas. Then I came out of the tank and Dave went in to brush out the dust.

We stopped at about quarter to twelve for some lunch. I then got geared up and went back inside the tank to do the fibre glass repairs. Dave poured some solvent into an old plastic paint pot and passed it into me with some brushes. It didn't take long for me to degrease the five patches that needed repair. I then started to lay up the fibre matting. Suddenly my air supply stopped and I dropped everything and put my head out of the manhole. Dave immediately pulled off my mask, and was yelling at a forklift truck driver whose wheels were on the airline. I think that I may have kicked over the pot of solvent inside the tank.

As I was already partly out of the tank I came out for a cup of tea, before going back on the job just before two o'clock.

Because the BA airline was damaged, I decided not to do any other work that day, and so I put on an ordinary disposable mask and nipped into the tank just to get my tools. Once in the tank I called Dave to switch on the light. Suddenly I was in a fireball and I just dived at the manhole. Dave helped pull me out, and soon Iain Wachem was there as well. I don't remember much more until I was in hospital ... .

WITNESS 3 – DAVID CARTER GENERAL LABOURER GRPM LTD

My name is David Carter. I have worked for GRPM for six months as a general labourer. I've spent over half my time working with Jimmy Henderson. I like working with him. He is really good at his job and takes the time to explain to youngsters like me about the secrets of the trade. Jimmy's also the one who has told me about health and safety, because we have to work in some dangerous places. The boss is pretty good too – he gives us extra danger money when things are really bad.

This is the first time that I've worked at Lochside Engineering. Jim and I travelled to work in the van together, so we arrived at about 8 o'clock this morning. The plant wasn't ready for us, so we got our gear from the trailer and got things ready near the acid tank. It looked pretty dark inside the tank and so I went to the workshop to see if I could borrow an inspection lamp and extension wire.

After breakfast, we got the permit and then we got started. Because of the grinding dust, Jimmy was wearing a B.A. mask supplied from our own portable compressor. He'd finished grinding by about 11:30 and so I put on a paper suit and dust mask and went inside the tank to brush up the dust.

After the lunch break, we went back to the tank at about quarter past 12, to start the repairs. Jimmy climbed into the tank to start the patch repairs and he'd asked me to get some solvent which I found in the van. I poured out about ½ litre of solvent from the special container into an old plastic paint pot to avoid metal sparks and so on, and then I passed the paint pot with some brushes to Jimmy inside the tank. I was aware that a fork lift truck was working nearby offloading some pallets from a trailer. I had already shouted at him once to be careful as he was reversing very close to our compressor.

Suddenly the truck came really close. I saw Jimmy struggling to get out of the tank. I think that the fork lift truck's wheels were on the air line. I yelled to the driver and then lunged at Jimmy to pull him out of the manhole. The fork lift truck wheel must have stopped his air supply.

I thought that Jimmy looked quite shaken up and so I suggested that we go and have a cup of tea. After tea, Jimmy went back to the tank and inspected the air hose – it was split. Jimmy said that it was no good and we wouldn't be able to do any more work today. We would have to go back to our depot for a spare. Then he said that he would go and get his torch from inside the tank. Jimmy put on a paper suit and climbed back inside the tank. I heard him call to me to switch on the light and as I did, I heard a sound like a "woooff" coming from the tank and saw a flame shooting out of the manhole. Jimmy was trying to get out of the tank. I dragged him out and smothered his burning clothes with an old ground sheet. By this time, Iain Wachem had arrived and was shouting orders and calling for an ambulance and the fire brigade... ... .

J3v. Photographs

Photographs for the case study can be mocked up from existing photographs or by using existing equipment as shown here. If using existing photographs, ensure that you have permission and are not breaching copyright.

Lochside Engineering Ltd
Boiler House Sulphuric Acid Tank

J3vi. Additional information – Work Specification (Quotation) from GRP Maintenance Ltd and a memo from the Lochside Engineering Works Engineer.

GRP MAINTENANCE LTD

SPECIFICATION: **Lining Repair – Sulphuric Acid Tank**

The vessel would be ground to a rough surface to provide good adhesion for the repair. This process will require extraction and for the operatives to wear full chemical suit protection. (Extraction to be provided by GRPM Ltd)

Lay-up

- One layer of 600g/m^2 chopped strand and glass mat
- One layer of synthetic tissue impregnated with Attac 741-02B
- Final topping layer of Gel coat with paraffin wax added in a 10% Styrene solution.

Curing

Initial curing at room temperature for 24hrs.

Post curing using warm air until a 30 hardness is achieved.

All the above to be provided for **£1780.00**

(one thousand seven hundred and eighty pounds)

Signed Date

C. Swift 23rd May

....

Proprietor

for GRP Maintenance Ltd

Unit 2A, Airport Approach Road, Dylle, Colvend DH7 6DW

Tel =+44 1778 340782

Lochside Engineering Company

MEMO

From: Jamie Watt, Works Engineer
To: Iain Harding, Manufacturing Director
Date: 19th March

Subject: SULPHURIC ACID STORAGE TANK

This tank was fabricated from Attac GRP in 1993, by the Balmoral Group.

In 2015 the inner surface had deteriorated in several areas about 30cm x 30cm, such that the gel coat had entirely degraded resulting in exposure and damage to the glass fibres. Records show that repairs were carried out by GRPM Ltd in 2016 and 2021, using the same specification detailed by GRPM in their recent quotation of 15th March

Without the repair scheduled for August this year, the life expectancy of the tank is no more than 5 years. With an effective repair, it is expected that the tank will have a life expectancy of at least a further 20 years, which would take it beyond the economic life of the rest of the boiler plant.

I would encourage you to approve an expenditure of £1780.00 to allow GRPM to affect this repair.

James Watt

3Jvii. Copy of procedure/instruction

When using procedures as a part of a case study, they need to be realistic. It is best to base them on one of your relevant existing procedures. It is quite common that health and safety procedures are quite long, and so it is useful to trim down the content of the procedure so that only that part that is relevant to the case study is provided, otherwise people will get bored with just reading reams of text and may miss the salient point. It is important, however, to customise the procedure to match the organisation names used in the case study in order to make the example more realistic. In the following image, the full case study procedure is not replicated in order to conserve space.

Lochside Engineering Company

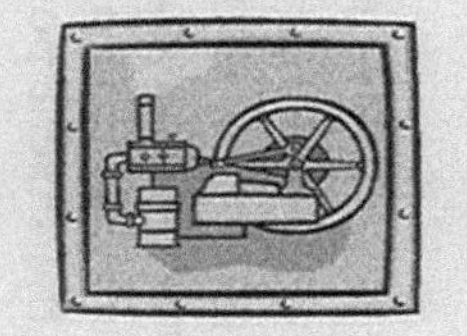

Procedure LE/S/003

ENTRY INTO CONFINED SPACES

Author: I. J. Harding

1. GENERAL

This procedure defines the requirements associated with clearance to enter and perform work in Confined Spaces on the Lochside Engineering Site. For general authorisation to perform work (including Hot Work) refer to procedure LE/S/001

2. DEFINITIONS

Authorised Person – as per the definition of Authorised Person in procedure LE/S/001, except that the authorisation shall specifically cover Confined Space Entries.

Confined Space – any place, including any chamber, tank, vat, silo, trench, pit, pipe, sewer. Flue or similar space in whiuch by virtue iof its enclosed nature, there arisise a reasonably foreseeable risk from:

- The presence of gases, liquids or solids which are flammable, toxic, asphyxiating, radioactive, hot or refrigerated, or
- Being liable to have the amount of available oxygen reduced to a dangerous level, or
- Having restricted means of entry

Confined Space Hazards – the hazards involved in inspecting, testing, cleaning, repairing or entering Confined Spaces, including:

- Asphyxiation
- Burning or scalding
- Electric Shock
- Fire
- Moving machinery
- Radiation
- Burial beneath solids
- Drowning
- Explosion
- Freezing

3Jviii. Learning from this exercise

There is a huge amount of learning available from this exercise, depending on what particular training the case study is being applied to.

Some conclusions when the exercise is applied to safe systems of work are:

a. There was no real specification or method defined for the job. GRPM's letter was purely a quotation.
b. Jimmy Henderson was proud of the fact that he knew how to do a good job of the repair. The secret was in the degreasing. Jimmy used a solvent to do this, but did not tell Lochside Engineering.
c. GRPM personnel did not understand the hazards of Acetone or that it was flammable.
d. The Confined Space Entry permit did not recognise that solvents were being introduced into the confined space.
e. Lochside Engineering failed to control neighbouring work, which resulted in the fork lift truck running over the long breather supply and cutting off air to the worker. It was this action which resulted in the tub of acetone being knocked over.
f. Jimmy Henderson has not been trained in Confined Space Entry. After the acetone was spilled, he went back into the tank without breathing apparatus, using only a dust mask. No air flammability test was done.
g. Although there was not much acetone in the old paint tub, when it was kicked over, it spread over the base of the tank, covering a very large area and suddenly increasing the surface area for acetone evaporation.
h. The source of ignition of the flammable atmosphere was the faulty light.

Conclusions when the Lochside Engineering Ltd exercise is used for Management of Change training:

a. The job method changed. (Acetone was introduced, but this was the tradesman's "secret" because this was how he achieved a "better quality" job than anyone else.)
b. The lighting arrangements changed. Initially there was no plan to use artificial lighting, but it became clear after the job started that lighting would be needed. No one realised that there could be flammable vapour present and therefore the lighting would need to be flameproof.
c. Working arrangements changed and allowed fork lift truck operations in the area where GRPM Ltd were working. This lack of control

over work which could affect the confined space entry initiated the incident when the fork lift truck damaged the air-line.

d. There was a change in the type of respiratory protection used. (Initially it was an air fed hood, but after the air line was damaged, this changed to a dust mask. However, this did not contribute directly to the incident.)
e. There was a change in understanding of the GRPM Ltd quotation. This was supplied purely as a cost bid, but was seen by Lochside Engineering's supervisor as a method statement.

KEY MESSAGE

Change is not limited to hardware changes – we need to control changes in peoples' behaviour/ways of working/organisation.

Section K

Automated PowerPoint exercise for emergency simulation

K1. Application

Emergency and crisis management training. One of the problems with emergency and crisis management is that most people get very little real opportunity to practice because thankfully significant accidents are very rare. It is very difficult to reproduce the stress and pressure that arises when a significant emergency arises.

To deal with this problem, any exercise needs to be able to recreate the sort of urgency and panic that can occur when things go badly wrong. A very effective way of doing this is to produce a real-time incident using a fully automated PowerPoint show. One of the rules of such an exercise is that the participants are not allowed to stop the automated show. This type of exercise is used during emergency management training once the basics have been taught.

K2. Location

This exercise is done in break-out rooms. Each break-out room must have a facility for projecting the PowerPoint images. This can be done by use of an LCD Projector and screen or by use of a large VDU display. Each break-out room must be provided with a laptop computer pre-loaded with the exercise files.

K3. Teams

The exercise will determine the size of the teams. In the case of the incident described, there would be approximately six team members, who would role play the emergency management team in action:

K4. Preparation

Setting up the PowerPoint exercise does take a significant effort, but once it is done, it can be used many times over. The experience of trainees almost

DOI: 10.1201/9781003342779-29

always is that they feel under considerable pressure, and are often not prepared for the pressure and unexpected side effects of managing an incident.

The emergency scenario should be story-boarded and then scripted before attempting to construct the PowerPoint presentation. The more generic the scenario, the more often the exercise can be used. I would recommend basing the incident on a fire, as this is a common incident in all sorts of industries and organisations.

The entire exercise is encapsulated in the automated PowerPoint presentation which usually will need to run continuously for about an hour. The exercise should be done in groups of four to six people. The initial slides must tell the participants that they are the management team of a particular organisation and that at some stage in the near future they will experience a major emergency. The slides tell them to get themselves prepared to deal with a sudden emergency by allocating roles, such as:

- Main Controller
- Incident Controller
- Heath, Safety and Environmental Specialist(s)
- Human Resources Manager
- Public Relations/Communications Manager

It is advisable, in this part of the exercise to provide background data that may or may not be required during the incident. This information is best provided on hyperlinks to such information as:

- What does the company do?
- What other companies/schools/old people's homes are nearby
- Material Safety Data Sheets, etc.

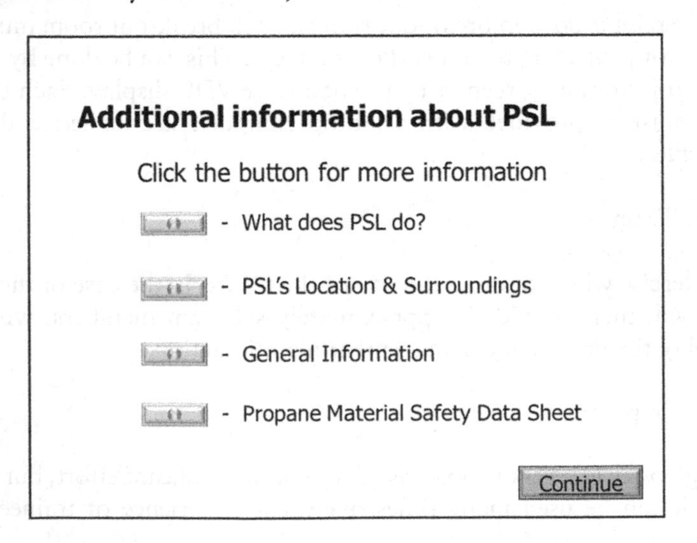

Example of Powerpoint slide showing information sources

The incident itself can be made up of a series of photographs or video clips. If possible, the scenario should include consequential effects such as the need to evacuate nearby buildings, traffic congestion, public interest, well-meaning helpers and un-associated medical emergencies like a heart attack among on-lookers.

Information about the development of the incident should be included electronically in the PowerPoint file. These can be audio recordings of telephone calls and radio (walkie-talkie) messages requesting help or reporting damage and can be from a mixture of people such as employees, members of the public, journalists, emergency services, regulatory bodies and local politicians. These audio recording must be scripted in advance and should use different people for each character. It is surprising how willing company employees are to record a small part of your script. The audible information can also be supplemented by examples of emails, text messages, faxes from various concerned parties. Later in the incident it is useful to increase the pressure on the emergency management team by simulating TV news reports, and media websites and social media chats. The pressure on the trainees is built up by periods of information overload in order to see how they respond and which information they react to and which they do not. The pressure is emphasised by having the incident carry a visible timer indicating how much time they have left to the end of the exercise.

An hour-long exercise typically requires about 80 or more individually animated slides. Sound and video recordings are added to the required slide by using the "Insert" tab on the PowerPoint toolbar and then selecting either "Video" or "Audio" from the media box on the top right-hand corner of the toolbar. Unlike standard presentations, the more animation and loud noises that can be introduced, the more realistic and fun that the exercise will be.

Examples of how some of the PowerPoint slides might appear.

Automatic advancing from one slide to the next is achieved by use of the "Transitions" tab on the PowerPoint toolbar. The time for the slide to display is set in the "Advance Slide" box on the far right of the toolbar. The time should always be greater than the time for any audio message embedded in that slide. Always remember to un-click the advance "On mouse click" box or the automation will not work.

It is recommended that audio messages such as telephone calls or radio messages are preceded with a normal telephone or radio call tone, and that the trainees have to take some action to respond to the phone ringing. This is usually to press a button on the screen in order to initiate the message.

To replicate reality, there should be some periods in the exercise where nothing new happens and other periods of frenetic activity.

When the incident itself is over and normality is restored, the final phase of the exercise is automatically introduced by the automated PowerPoint presentation and should describe what the trainees have to do to complete the exercise.

K5. Other supporting materials

In addition to the development of the automated PowerPoint exercise, the trainee emergency management team should be presented with facilities to quickly record what is happening (an "Events Log"), a means of recording incoming information (some form of message sheet) and also an "Actions Log". In the training exercise the Events Log and the Action Log are best provided as large poster-sized sheets held on a flip-chart stand, so that all the team can see what is being written. The message sheets are best provided at A4/foolscap size.

Example of an Events Log sheet.

EVENT LOG **SHEET A**

Log Maintenance Responsibility .... Date ....

Seq No	Start Time	Event	Finish Time	Signed

Example of an Action Log

ACTIONS LOG **SHEET H**

Log Maintenance Responsibility .... Date ....

Number	Action	To be done by	Time raised	Priority	What was done?	By whom	Time done	Complete

Example of an Incoming Message Form

INCOMING MESSAGE FORM			
Date		Message Number	
Time		Message To	
Callers Name		Name	
Callers Location		Location	
TEL NO		Named person notified	

PRIORITY RATING (Circle Priority)	**1**	**2**	**3**

MESSAGE

ACTION TAKEN

Seen and signed by:			
Main Controller		Incident Controller	
Communications Co-ordinator		EHS Co-ordinator	
Technical Co-ordinator			

K6. Running the exercise

The exercise is best run with the actual team who would be responsible for managing a major incident or crisis and should be used after all the preliminary explanation training has been done.

I find that it helps the realism to have the display replicated from a computer screen onto either an LCD projection or a very large VDU. There should be no need for detailed introduction and explanation of the exercise as the automated PowerPoint should do that. However, it is absolutely essential that the automated PowerPoint has been tested to ensure that it runs continuously without a hitch. The tutor will need to ensure by observation that each of the trainees is fulfilling their role and that they are tacking decisions and identifying actions. One of the realistic features of this exercise is that there can be a tendency for trainees to get cognitive overload and freeze. In these situations, the tutor may need to prompt the trainee.

Part of the exercise is to create a feeling of lack of preparedness and being out of control, as this will encourage them later in their training to address some of these issues.

A limitation of the exercise is that with the standard approach then the trainees cannot interrogate phone messages and information provided. However, it is possible to set up a series of telephones where the respondent will role play to the trainees and provide information relevant to the exercise, including calling the emergency services. Unfortunately, this adds very significantly to the complexity of the exercise and requires a large number of additional tutors/helpers to be available and fully briefed.

K7. Feedback

There are two stages to the feedback stage of this exercise. Initially the trainee group themselves should be asked to identify:

- Did you feel in control?
- What did the team do well?
- What would you do differently next time?

The tutor will then provide his/her feedback on what went well, and what could have been done differently.

K8. Mock Press briefing

The final stage of the practical exercise is to carry out a mock press interview. Usually, the tutor will take the role of a rather aggressive TV reporter and will interview two or three members of the management team about what happened in the incident that they have just been managing. It is useful to video record this interview as it adds a further feeling of realism and pressure. The video recording may or may not be played back, depending on the learning opportunities and the individual trainee's sensitivities. The key message from this final stage of the emergency exercise is – *make sure that you have someone who is appointed to face the press and that they are properly "media trained"!*

Section L

Competitions – Trainees' presentations

L1. Applications

Making a training session a bit competitive adds hugely to the enjoyment of the topic and if the trainees are enjoying it, they will be engaged and learning. Competitions can be used to confirm levels of understanding either of what trainees already know, or to confirm what they have learned during part of the training. A quick and fun way of doing this is to use quizzes (see the next Resource Section M).

So often training is done by a trainer who is in "telling" mode. In this situation the trainer is seen to be "all-knowing" and the trainees apparently lacking in relevant knowledge. This is palpably not the case. Everyone has a wealth of knowledge on different subjects. It is very powerful to use some of the trainee's knowledge and share it with each other. This approach recognises and values the knowledge that trainees already have. I have regularly used this approach to share knowledge where trainees are coming from different departments, locations or countries. Its main application is in the sharing of health and safety knowledge so that we can all learn from others experiences without having to learn ourselves the hard way.

L2. Preparation

Before the training begins, each trainee is asked to come prepared to talk about some experience that he or she has had that they think will be of benefit to the others. They are told that they will be limited to 5 minutes to explain their message, and that they can use a PowerPoint presentation of no more than five slides, or they can use flipcharts or whatever other technique they choose to get their message across. One trainee actually sang a song – everyone remembered that one!

The presentations are run competitively and the judges are their colleagues and not the tutor. After each presentation all the other trainees will record a score made up of:

DOI: 10.1201/9781003342779-30

Level of interest	Score from 1 to 5
Quality of presentation	Score from 1 to 5

Naturally, trainees cannot vote for their own presentation! The best overall score is awarded a suitable trophy at the end of the training session. This is a good opportunity to involve a senior manager to come and make the presentation as it adds to the competitiveness and allows the senior manager to demonstrate his own personal commitment to health and safety. I tend to present a trophy that is durable and that is in some way prestigious. A small and inexpensive engraved glass trophy is ideal because it ends up on a departmental or office shelf and becomes a permanent reminder of the training and is often a discussion point for many years after the training has passed.

Presentation by Senior Manager at the end of a Delegate's Presentation Competition

L3. Running the competition

It is useful to have any trainees' presentation electronic files already preloaded onto the tutor's computer before the training starts to avoid unnecessary delays. If there are a large number of trainees, then it is advisable to split the trainees' presentations into groups of no more than three. I would tend to do these at the start or end of a day as a bit of variety.

The challenge for the tutor in this type of session is ensuring that the trainees do not over-run their time. Although the trainees will have been told that they have 5 minutes, I always allow a one-minute over-run. A timer board is used to let the speaker know how much time is left. I would normally ask one of the other trainees to operate the timer board as it adds to their interest and engagement. Astonishingly, the presenter tends not to notice the time although their fellow trainees will be watching it carefully. If the presenter exceeds 7 minutes, then they must stop. A fun way of bringing an over-run to an end is to use a recording of some loud event – I like to use Tchaikovsky's 1812 Overture to sound when the time is up. To make this automatic, this is done by recording 7 minutes of no sound followed by the 1812 Overture. Just by clicking on the computer audio file, leaves the tutor free to give full attention to the presentation. I have never yet had a series of trainees' presentations where Mr Tchaikovsky and his band haven't struck up at least once – much to the audience's delight!

When recording scores. it is beneficial to have a simple score sheet for each trainee. If the presentations are being run over a number of days, the trainees will need to be reminded to keep their scores up-to-date.

When compiling the trainees' feedback at the end of training courses I find that it is very common for the trainees to feedback that the trainees' presentations were the most enjoyable and useful part of the training session. This is almost certainly because they appreciate the fact that their own knowledge and experience is being valued.

Section M

Quizzes

M1. Applications

Quizzes can be used for almost any training subject. They can be used as a fun way to check understanding, reinforce messages and carry out validation.

M2. Location

Done in the main training room

M3. Teams

Done in teams of three in order to get discussion between the team members

M4. Timing

A quiz of 10 PowerPoint displayed questions takes about 10 minutes, including displaying the answers.

M5. Preparation

The quiz should be done in a way that the trainees find to be fun, and does not remind them of school tests. Running the quiz in small teams of two or three makes it less threatening and making the format of the quiz like a pub quiz or "Who Wants to be a millionaire" with a small prize (chocolate bar) for the winners adds to the ambience and encourages participation.

The least threatening type of quiz is one of multiple choice. Once a framework has been produced, the questions and answers can be easily changed to cover different subjects or groups. My preference is to run the quiz as a PowerPoint presentation, so that there is some anticipation builds up among the trainees. The format that I use is shown below:

DOI: 10.1201/9781003342779-31

Welcome to the

Confined Space Entry-Quiz

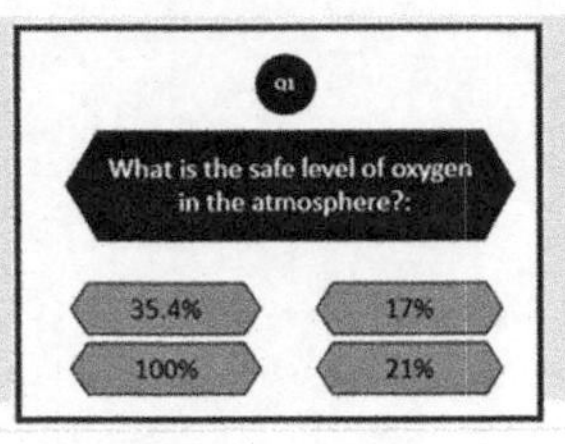

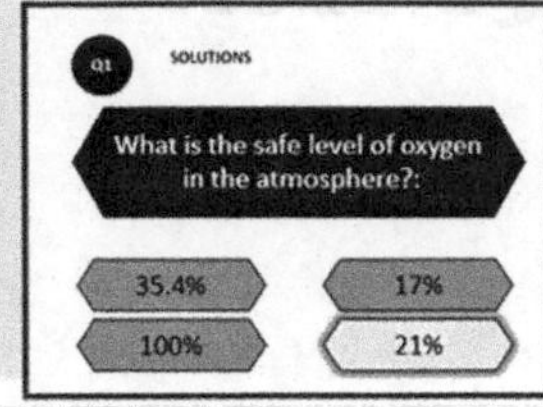

Each team should have a pre-prepared answer sheet. To aid memory do not ask the trainees to record answers "a, b, c or d" but ask them to write down the correct answer, which in the case of the above example would be "21%". Typically, I would suggest that the quiz has about 10 questions. It is best to ask the questions first before going to the answers. To encourage participation, ask the teams to call out the answers that they have got before revealing the correct answer. Anticipation is raised if the questions and answers are animated on the PowerPoint slide.

Quizzes can be used not only to test people's learning but also as a learning method in its own right. This is particularly useful when teaching some detailed elements of legislation. In this application (see example below), the tutor does not expect the trainees to know the right answer, but to take an inspired guess. When the answer is given, the fact that it can come as a surprise will make it more memorable. However, this application of quizzes should be used very sparingly.

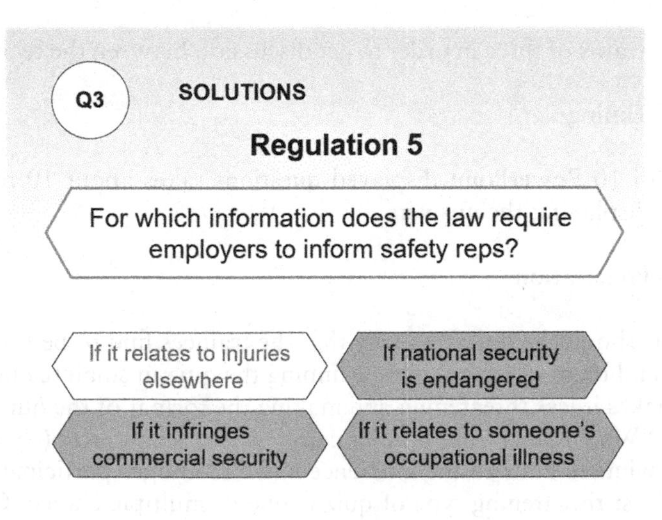

Section N

Puzzles (communications exercise)

N1. Application

This puzzle exercise is used primarily to explore the problems that can occur in communications. It applies especially to verbal communications over the phone or radio where the individuals involved are not in visual contact. This represents a large proportion of communications at work

N2. Location

The communications exercise is run in break-out rooms, corridors and suitable places that are close to the main training room. The key message is that no two teams should be within earshot of one another.

N3. Teams

The exercise uses groups of three trainees. Two of the trainees take part in the exercise and the role of the third is to observe and provide feedback.

N4. Running the activity

The two trainees are asked to hold a conversation and are positioned so that they can hear one another but cannot directly see each other (they are seated back-to-back). One trainee is designated as the "Leader" and the other as the "Follower". The leader is given an image of a set of coloured shapes which are displayed in a final image. The follower has the actual set of coloured cards which could make up the final image. By verbal communication alone, the Leader has to get the Follower to put together the loose shapes to form the finished image that the Leader has been given.

DOI: 10.1201/9781003342779-32

Arrangement of main participants in the Communications exercise

Preparation: Make a set of coloured pieces that can be put together to form the final image. Place these together and scan them and print the finished image for the use of the leader. Place the finished print in a large envelope, so that it cannot be seen until the exercise is ready to start.

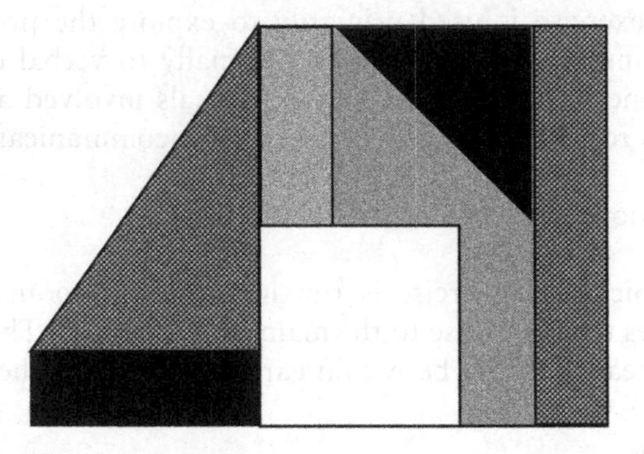

Normally a number of these exercises will be run in parallel. It is advantageous that the coloured pieces that the Follower is using do not match the colours on the Leader's image. Part of the exercise is to establish quickly that the only thing that matters is the shape of the pieces and not the colours.

Timing: The exercise takes about 20 minutes and should be stopped once one team has completed the exercise correctly. A further 5 minutes is taken for the observers to feedback their comments.

Section O

Mock scenarios

O1. Application

Suitable for safe systems of work, energy isolation plans, confined space entry and scaffold standards training exercises.

Mock scenarios can be used where there would otherwise be a need to have access to potentially hazardous plants and equipment in order to complete effective training.

SCENARIO NO. 1 – SAFE SYSTEMS OF WORK/ENERGY ISOLATION PLANS/PERMITS TO WORK

Preparation

The scenario can make use of idle, redundant or re-purposed equipment. The example shown here is used during training in a large hotel and conference centre and makes use of their water and space heating plant. In this case the exercise is intended to demonstrate the preparation of a hazardous hot oil system for urgent maintenance, by creating a potentially hazardous fictitious scenario in realistic surroundings, using relatively low-risk water circulating equipment. The central heating plant used is a part of a large wood pellet burning boiler system connected via pumps, heat exchangers and valves to storage tanks. In order to increase the realism, all the equipment is relabelled with temporary hot oil labels and hazard symbols.

DOI: 10.1201/9781003342779-33

A simple process line diagram is provided, together with a set of wired tags to allow the trainees to clearly identify which locations are to be isolated or blinded with spade plates.

A portable replica of a leak point on the floor which has been caught in a small drip tray filled with absorbent granules all adds to the realism.

Running the exercise: The exercise is most beneficial if run in groups of two or three trainees together, in order for them all to be involved and to get good discussion going.

The Task

Exercise Safe Systems of Work – Preparation for maintenance – Trainee's Briefing Note

Location – the Hotel's main plant room

Your Group's task

As a part of the job preparation, you are to prepare an energy isolation plan for the job described below. You are provided with the following:

1. *A Material Safety data sheet for Dowtherm*
2. *A simplified line-diagram*
3. *Energy Isolation proformas*
4. *A selection of wired tags for on-site identification*
5. *Simulation of the inside of the electrical isolation cabinet*

You will have a total of 20 minutes to complete your task. When the hooter sounds you must move onto Exercise 2.

Please remember that although you will need to attach tags to the plant, the preparation of the energy isolation plan does not involve you in carrying out any isolations or removal of any lagging or other parts of the plant.

The job to be done

The system that you will be working on is a closed loop heating system linked to a Biomass burning heater. The process fluid is a heat transfer oil called "Dowtherm". The Dowtherm is heated to 375DegC. A leak has been spotted coming from under the lagging on a pipe at the front of the Dowtherm storage Tank.. Depending on the nature of the leak, repair work may need a further permit to work, or management of change assessment. However, your isolation plan should be adequate for both the investigation and repair of the leak.

Your isolation plan should identify and mark up (with tags) all the process and energy isolations required to make the job safe. Process isolations should include valved isolations, and slip plate/blind positions and physical disconnections if necessary.

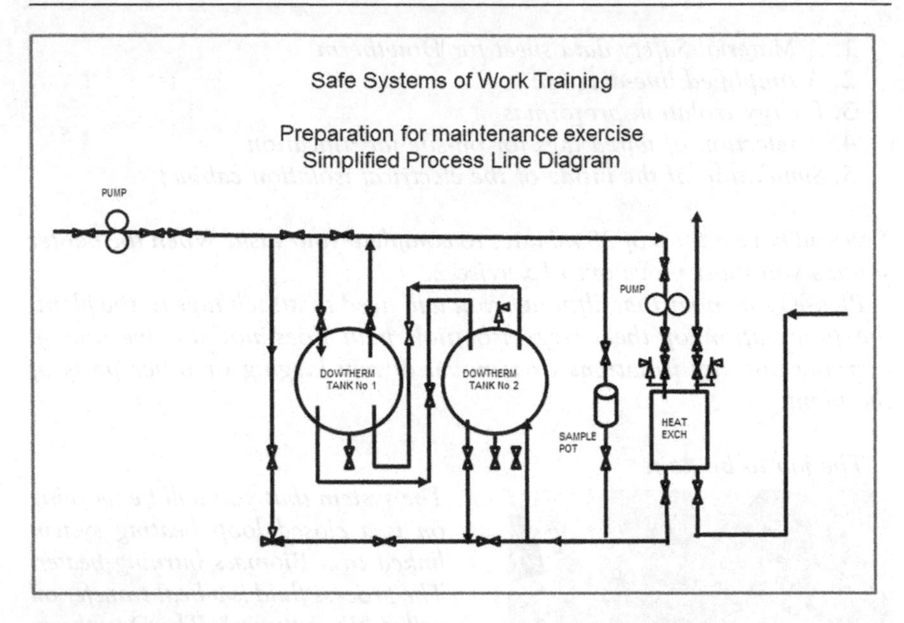

For the purpose of this exercise, electrical isolations will be assumed to be carried out by placing a sticker on the relevant isolator on the photograph of the isolations of the drive cabinet shown below:

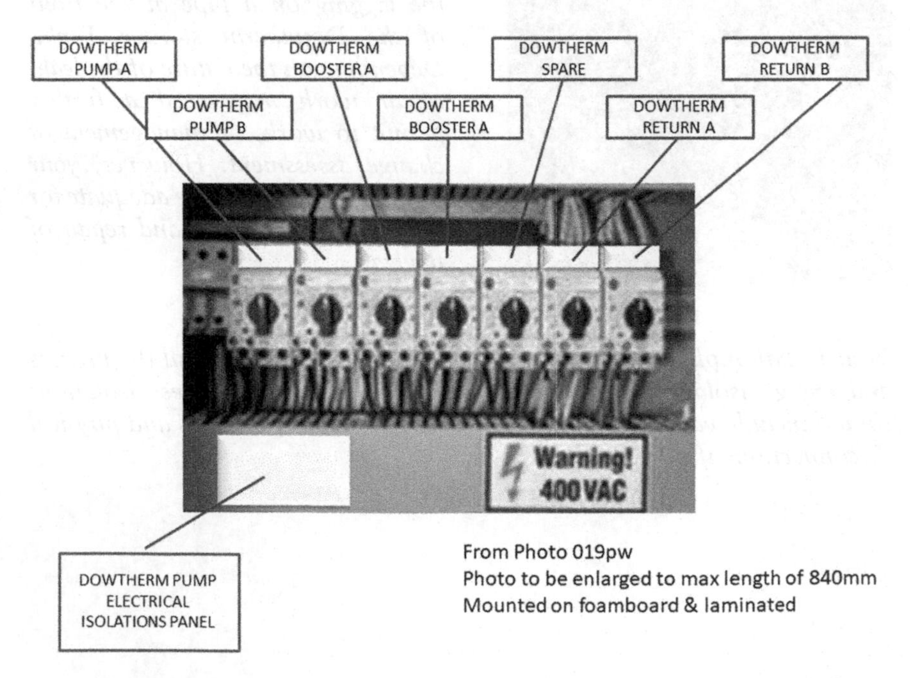

Timing: Allow 25 minutes to complete the task, followed by group feedback. Remember to allow for travelling time to and from the "Mock-up".

Example of Tutor's Briefing Note
– Safe Systems of Work – Preparation for maintenance exercise/Energy Isolation Plan

Information provided:

- *Tank and pump labels (these should be in place before starting)*
- *Trainee's brief (Enough copies for one/trainee)*
- *Dowtherm Material Safety Data Sheet (printed two-sided)*
- *Dowtherm Circuit Diagram*
- *Example of an Energy Isolation Plan*
- *Blank Energy Isolation Plan forms*
- *Wired tags (20)*
- *Clipboard*
- *Post-it notes for use on the electrical isolations photograph. (Note: the photograph should be large enough to easily accommodate the slickers.)*

The tutor should start by explaining the Dowtherm heating circuit which is heated by the adjacent Biomass heater.

Indicate the locations of the Dowtherm tanks 1 and 2 and the three sets of pumps.

Point out the location of the leak on Tank No.1 inlet. The leak is under the lagging, but it is not clear at the moment which side of the valve it is. There is a simplified MSDS. Don't let them all read it – it only needs one person to read it and report their findings to the others. The key point is that the Dowtherm is very close to its autoignition temperature, so any serious leakage would form a vapour cloud and might explode.

To examine the leak, it is necessary to empty tank 1. You can assume that there is sufficient ullage in tank 2 to take the contents. To achieve this, the tutor will be looking for the trainees to:

1. *Stop the process of heating and shutdown the Biomass heater*
2. *Shutdown the Dowtherm pumps A and B*
3. *Close the Dowtherm pumps delivery valve and Tank 1 inlet*
4. *Shut the Tank 2 delivery valve*
5. *Use the Booster pumps to pump from tank 1 via the heat exchanger to tank 2*
6. *When transfer is complete isolate the three sets of pumps (process and electrical)*
7. *Check crossover line between the two tanks is isolated*
8. *Check that the two tank by-pass valves are closed*
9. *Check if leak has stopped*
10. *Check that temperature of tank 1 has dropped to well below the autoignition temperature*

11. *Remove blank on Tank 1 drain valve and check that tank is empty and de-pressured*
12. *Remove insulation and check for leaks on either side of valve*

The best way for the delegates to do this is that they put each of their planned actions on post-it notes until they have got the sequence right and then they can list the correct sequence on the EIP proforma and after that they can go out and put tags on the system before completing the EIP "Ref" column.

Tags are available in the pack (note that tag 6 and tag 9 are intentionally confusing. It is important to get them to recognise that tag numbers must be unique as there could be old tags out on the equipment.

Electrical isolations are deemed to be complete by putting a small post-it note on the appropriate isolator on the switchgear photo.

SCENARIO NO. 2 – CONFINED SPACE ENTRY

Preparation: The scenario can make use of idle, redundant or re-purposed equipment. The advantage of using a safe place and mock-up for training is that it eliminates the need for taking equipment out of use just for training purposes and the subsequent need for making line break isolations and air tests and the associated potential hazards. The preparation requires the foresight to take detailed photographs of the internal feature of the equipment that will be featured in the training mock-up exercise. These photographs should be reproduced in large poster-sized format so that detail can be easily seen. In the exercise description that follows, the mock-up equipment to be used for the confined space entry is a fired boiler.

Running the exercise: The exercise is most beneficial if run in groups of two or three trainees together, in order for them all to be involved and to get good discussion going.

The Task

Preparation of Confined Space Entry certificate

Location – the Wood burning Boiler

The boiler produces hot water for heating and domestic hot water. The photograph shows the main access to the combustion chamber.

Your Group's task

Your maintenance team need to carry out repairs to the damaged refractory lining (brickwork) at the back of the boiler's combustion chamber, known as the "quarl". The quarl is the circular-shaped brickwork around the burner point, which has been seriously eroded through use. The boiler has water, air, ignition gas and biomass (wood chippings) supplies. The task entails full entry into the boiler via the front-hinged igniter door. Power tools may be required to break out the old brickwork.

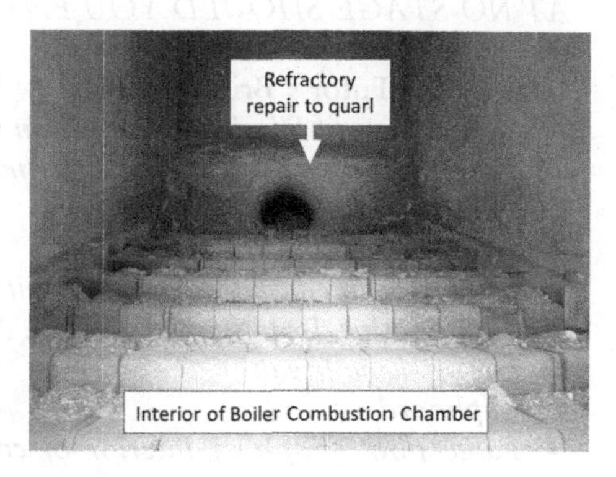

View inside combustion chamber

You should prepare a confined space entry certificate for the task. You should survey the boiler and identify and record what isolations are required to allow for the work to be carried out safely.

You should identify on the confined space entry permit form provided:

1. *Isolation requirements*
2. *Oxygen levels and flammable gases.*
3. *Other risks*
4. *Standby (Guard) requirements*
5. *The elements of the rescue plan*

For the purpose of this exercise, electrical isolations will be assumed to be carried out by placing a sticker on the photograph of the isolations on the relevant drive cabinet.

Resources provided

1. *Large scale photographs of the boiler interior*
2. *Large photograph of the electrical isolations*
3. *Oxygen concentration information*
4. *Confined Space Entry Form*
5. *Tags (labels)*

You will have a total of 25 minutes to identify the necessary physical and electrical isolation points and to complete the confined space entry certificate.

AT NO STAGE SHOULD YOU ENTER THE BOILER!

Example of Tutor's Briefing Note
– Safe Systems of Work – Preparation Confined Space Entry Permit
This exercise is best run with a tutor/mentor in attendance throughout.
Information provided:

- *Blank Confined Space Entry Permit forms*
- *Self-adhesive Isolation labels (e.g. "Valve isolation": "Spade isolation": "Line break")*
- *Clipboard*
- *Large photograph of interior of confined space, showing intended work area*
- *Large photograph of inside of electrical isolations cabinet*
- *Post-it notes for use on the electrical isolations photograph. (Note: the photograph should be large enough to easily accommodate the slickers.)*
- *Photograph of a reading of gas detector*

Start by explaining to the trainees the main aspects of the boiler system so far as it relates to the confined space entry.
Indicate to the trainees that they need to

A. *identify all the possible hazards including those introduced by the task itself*
B. *identify the physical isolation points and the type of isolation required*
C. *identify the electrical isolations required on the electrical isolations photograph*
D. *identify what arrangements will be made for air circulation*
E. *identify what arrangements will be made for lighting*
F. *identify and detail a rescue plan*

Photograph of a trainee simulating electrical isolations on the isolation photograph

Prior to preparing the confined space permit the trainees are reminded of the need to take air tests. In the mock-up scenario, this information is provided to them in a picture of a gas analyser which has had unsatisfactory readings superimposed upon it.

The trainees need to recognise that the oxygen level shown on the analyser picture is unacceptable. They must investigate the cause of low oxygen and request a further gas test. At this stage, if they have taken some (theoretical) action to increase ventilation rates, they are given a second picture of a gas analyser showing acceptable (21%) oxygen levels' The exercise is completed by the correct completion of the confined space permit.

BIOMASS BOILER INITIAL AIR TEST RESULTS

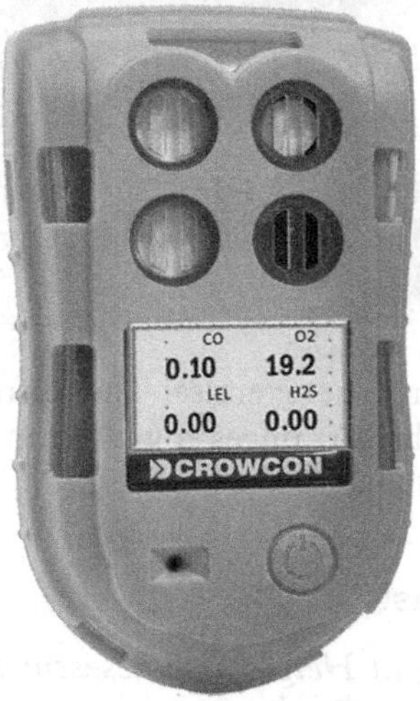

CO 0.10 O2 19.2 LEL 0.00 H2S 0.00 CROWCON

Photo by kind permission of Crowcon Detection Instruments Ltd

SCENARIO NO. 3 – SCAFFOLDING STANDARDS/WORKING AT HEIGHTS

This exercise is intended for the use of scaffold users, and not professional scaffolders. A lot of people in the workplace do not recognise that a scaffold structure and platform can be technically perfect but that it is not a safe place of work because it is not quite in the right place or at the right height. It is important that scaffold users have a basic knowledge of what constitutes a good and safe scaffold.

It is relatively easy to set up scaffold structures to demonstrate scaffolding faults, but if the use of these mock-ups is required for several training courses which do not run concurrently, these real mock-ups can be expensive, very time consuming and expose scaffolders to unnecessary risk in building platforms that will not be used for their main purpose.

This exercise uses the principle of only producing the real scaffold with its "designed in faults" on one occasion and then using video film and large photographs to replicate its use as exercises in multiple training events.

Using large photographs of scaffold minimises risk to trainees

Running the exercise: The exercise is most beneficial if run in groups of two or three trainees together, in order for them all to be involved and to get good discussion going.

The Task

Working at Height risk assessment/scaffolding standards

Trainee's Task

There is a need to fit an additional fibreglass roof-light in the roof of the store. The position of the roof panel to be removed is marked on site with an "X" and shown in the photograph as dotted lines. A scaffold has been erected to provide access for the work. Using the large photographs, video film and your own knowledge you should apply the six steps of risk assessment to identify whether the risk is tolerable and whether there are any additional precautions that you need to take before a Permit to Work is issued.

Location of new skylight

Remember the six steps of Risk management. You should:

1. ***Identify the hazards***
2. ***Identify who/what might be harmed***
3. ***Assess the risks***
4. ***Decide what extra controls are necessary***
5. ***Record/communicate findings***
6. ***Review the assessment** (Note: You will not be required to carry out a review for this exercise)*

Example of Tutor's Briefing Note

– Safe Systems of Work – Working at Height/Scaffolding Standards

Information provided:

1. *3x A1 size photographs of the scaffold access to the roof. (These will be already set up on three flip chart stands.)*
2. *Video file showing details of the access scaffold. Showing time = $1^1/_2$ minutes (Note: You will need to provide a fully charged laptop loaded with the video as there is no power available.)*
3. *Exercise briefs (one per trainee)*

Time allowed is 25 minutes

The working at height risk assessment

The task is identified on the trainees' briefing card. The sheet to be removed and replaced with a translucent fibre glass panel is shown highlighted in whitechain dot. The delegates have not been told that the technique for removing the bolts in the old roofing sheet is not by burning, but by air-arc gouging (effectively using an electric welding torch with a special electrode which acts in the same way as oxy-acetylene burning). The clue that the delegates have been given is the welding set on the scaffold. This would give rise to sparks and there is lots of combustible biomass fuel around and other combustible items stored.

Step 1 – Identify the Hazards (Examples)

- *Working at heights*
- *Hot work with combustibles*
- *Delegates may identify asbestos cement roofing sheets – these sheets are non-asbestos*
- *Mobile equipment entering building*
- *People*
- *Condition of scaffold*

Examples of scaffold faults

1. *Presence of person on photo shows that the scaffold platform has been set too low for this task*
2. *The top end of the panel is well outside the plan area of the scaffold. It should have been wider*
3. *Missing diagonal brace on ladder side of scaffold*
4. *Diagonal brace pole sticking out and creating trip hazard*
5. *Missing floor-plate*
6. *Missing lower handrail*
7. *No ladder gate*
8. *Ladder should extend at least 1 metre above the scaffold platform*
9. *Transom under scaffold is spaced >1.5metres from the next transom*
10. *Missing toe board on fuel bunker side*
11. *No scafftag or visible inspection certificate*
12. Should not use extending ladder

Step 2 – who or what could be harmed (Examples)

- *Person doing work (could fall)*
- *Person doing work (burn or arc-eye from air arc gouging)*
- *Dust in eye*
- *People not associated with the job. E.g. gardener or is a fuel delivery expected?*
- *Fire in fuel/stored items leading to people/building damage. Damage to the boiler system would seriously affect the hotel's ability to trade.*

Step 3 – Identify the risks

The risk matrix shows that the risk is currently not tolerable. There is a high risk of falling from height and also of fire

Step 4 – What extra controls are necessary (Examples)

Is the job really necessary – the building is already well lit by skylights. However, if it is assumed that it is required then:

1. *Correct all the faults in the scaffold shown in the list 1–9 above*
2. *Use cold cutting techniques for the removal of the bolts in the sheet*
3. *Control access into the fuel bunker bay (lock roller door or put a barrier across the opening)*

Step 5

Record your information. (Normally this would go onto the permit – but in this case if time allows it should go onto a flip chart.)

At the end of the exercise issue a copy of the HSE Information sheet (general access scaffolds and ladders CIS No 49) to all delegates.

Timings: The exercise takes 20 minutes for each team to complete.

Section P

Personal protective equipment exercise

P1. Applications

Practical exercise for the identification of personal protective equipment (PPE) and respiratory protective equipment (RPE) for chemical hazards

P2i. Teams

Run in teams of 4–6 trainees

P2ii. Location

Usually run in a break-out room

P2iii. Timing

Exercise takes approximately 20 minutes. Feedback can take an extra 10 minutes.

P3. Preparation

The exercise requires the availability of a selection of PPE that should include for this example among other things:

- Hard hat
- Dust mask
- Overalls
- Dustproof gloves
- Safety boots

The additional items should be included to make the trainees think. For example, it might also include things such as bump cap, balaclava, goggles,

DOI: 10.1201/9781003342779-34

light eye protection, hearing protection, surgical gloves, short-sleeved worktop, leather apron and chemical suit.

The trainees also require copies of the up-to-date Material Safety Data Sheet and their remit.

P4. Running the exercise

If there are several syndicate groups involved, then it is best to allocate a break-out room to each for this exercise. The use of hydrated lime in the example below should be replaced by a substance that is relevant to the trainees present.

Task remit
Hydrated lime is used in the building industry for surface coatings. Your group are working on renovating an old building and need to mix and apply hydrated lime to the external walls. Using the MSDS provided (for Hydrated Lime) decide what precautions need to be taken and what personal protective equipment will be required to control the risks during

a. *The lime preparation activity*
b. *The lime rendering activity*

The lime is supplied in 25 kg polythene lined sacks and is mixed with water in open-topped tubs using an electric paddle mixer.

When you have assessed the risk and decided what PPE should be worn, the group should choose suitable PPE from the selection provided to ensure that the hazards identified in the MSDS are controlled. One member of the group will be required to wear the specified PPE when returning to the main training room and each syndicate group will be asked briefly to explain what PPE they chose and why.

If there are several groups involved, it is best to give each group a different hazard.

Section Q

Noise simulation exercises

Q1. Applications

These exercises are used to help trainees get a good understanding of the decibel scale of noise in a controlled environment where their hearing will not be harmed.

Exercise 1: Quick noise demonstration to large groups.

Because the noise measurements are logarithmic, few people really understand what a noise level of 80dbA really sounds like. I have heard many workers say, "but the noise level was only 3 decibels above the limit" without realising that means that it was twice as loud!

Many factories are inherently noisy places. When I started my working life in the steel industry, it was a matter of pride to be deaf, because that meant that you had worked there a long time and therefore must be very experienced! The problem is that people don't understand noise levels or the meaning of time-weighted averages.

If the training is being done with a large group, then a very simple way of demonstrating noise in a memorable way is to get the audience to make as much noise as they can, by clapping, shouting and stamping their feet. To give immediate feedback they need to be able to see the effect of their efforts. This can be done by projecting the readout of the sound meter onto a large screen – either directly or through the use of a video camera.

This exercise takes about 2 minutes to run and is very memorable.

Exercise 2: Noise appreciation exercise (Syndicate Group exercise)

This exercise is designed for groups of 4–6 trainees and is intended to be an appreciation of what are common noise levels at work and what could be done to reduce them. The exercise involves six separate tasks and the time taken for the exercise is 20–30 minutes.

DOI: 10.1201/9781003342779-35

Preparation: The exercise requires about five audio (or video with audio) recordings of a range of different workplaces ranging from noisy to not-so-noisy. Include some examples where the result is surprising (e.g. commercial kitchens or vehicle cabs). The sound source (speakers) should be set to maximum volume. Tape floor markers should be set up 2 metres and 4 metres from the sound source.

Remember to make sure you know how to operate the sound meter before the exercise starts!

Equipment required:

- Noise meter
- Laptop (with noise recordings on file) with external mains speakers on maximum.
- A1 chart (Noise Exercise Tasks 1 and 2) on flipchart stand plus marker pens
- Noise measurement points 2 and 4 metres from speakers

Running the exercise:

TASK 1
Ask the group to consider the noise levels in the following situations (adapt these situations to suit your situation)

- *External areas of a continuous process plant*
- *A commercial kitchen preparing food*
- *A busy office*
- *A laboratory*
- *A large mechanical press*

Ask the group to guess where the five noise sources on their list would be on the Noise Exercise chart shown below. Get them to mark each of these situations in the 1st guess column on the chart shown below. Don't let them linger on this too long (3 minutes max).

As a guide:

- If normal conversation is possible at a distance of 2 metres (i.e. not shouting to be clearly understood) the noise level is probably less than 80 dB(A)
- If is it necessary to shout at a distance of 2 metres to be heard? The noise level is probably >85bB(A)
- If is it necessary to shout at a distance of 1 metre to be heard? The noise level is probably >90 dB(A)

Noise Exercise Chart (ideally this should be drawn in advance on a large flipchart)

	dBA	1st Guess (before hearing recordings)	2nd Guess (after hearing recordings)	1st Measurement (at 2 metres)	2nd Measurement (at 4 metres)
Threshold of Pain →	140				
	130				
	120				
	110				
	100				
	90				
	80				
	70				
	60				
	50				
	40				
	30				
	20				
	10				
Threshold of Hearing →	0				

TASK 2

Now ask the group to play the noise recordings to revise their guesses in the 2nd column of the chart. Discuss briefly with them why they may have changed their minds.

TASK 3

With the sound meter in a fixed position 2 metres from the sound source, measure the noise levels of each of these situations using the noise meter provided and record the results in the 3rd column on the chart.

Ask the group which situations that you heard could pose a threat to your hearing?

TASK 4

Consider the loudest noise source and identify how long you could be exposed to that noise in an 8-hour working shift, before you are at risk of hearing damage.

TASK 5

Assume that your staff need to work in that loudest noise situation for most of their working time. Using the hierarchy of controls to identify what actions could be taken to ensure that individual noise exposure is reduced and their hearing is protected.

Hierarchy of Controls:

1. *Elimination*
2. *Substitution*
3. *Engineering Controls*
4. *Systems of Work*
5. *Behavioural Modification (Training and Auditing)*
6. *Personal Protective Equipment*

Produce your responses to each task on a flip chart

TASK 6
If time allows reposition the sound metre at the 4metre mark and repeats the sound measurements. Observe how distance reduces the sound level. Remind the trainees that because the dBA scale is logarithmic, small changes in the decibel level can reduce the noise exposure significantly (approx. 3 dBA reduction equates to a halving of sound exposure).

Timing: The exercise takes approx. 30 minutes

Section R

Manual handling assessment exercise

R1. Applications

Manual handling means **transporting or supporting a load solely by hand or bodily force.** It includes lifting, putting down, pushing, pulling, carrying or moving loads. It can be used to identify when human effort alone is likely to lead to some level of musculoskeletal harm. This exercise can be used to test trainees understanding of carrying out a manual handling assessment or it can be used as a tutor lead training exercise in small groups.

R2. Preparation

The exercise is intended to demonstrate how to carry out a manual handling assessment and to show that harm arising from manual handling is not always obvious. The exercise will get a small group of trainees to go through a progressive exercise to assess the likely risk from handling commonly found loads at work.

The preparation requires the printing of a large wallchart poster (typically A0 sized) showing how to carry out the manual handling assessment. Five example weights should be produced.

The weights are:

Item A. Box file full of papers
Item B. A briefcase
Item C. A sack of potatoes
Item D. A box of copier paper
Item E. A bag of groceries

At least one of these examples should be created to be misleading weights. The 25 kg sack of potatoes should be filled with polystyrene packaging or empty plastic drinks bottles so that its weight is not 25 kg, but about ½ kg. Likewise, the briefcase can be filled with heavy books to be heavier than normal. There should be sufficient hand-held luggage scales to allow each trainee to weigh accurately weigh each load. Each load must have a lifting

DOI: 10.1201/9781003342779-36

loop attached to allow the scales to lift that item. The reason for some deception in the weights of the loads is to demonstrate that physical appearance of an item is not necessarily a good guide to its weight.

The exercise needs a number of handouts to be prepared in advance the details of which are shown in the next section

The wall poster content is as shown here:

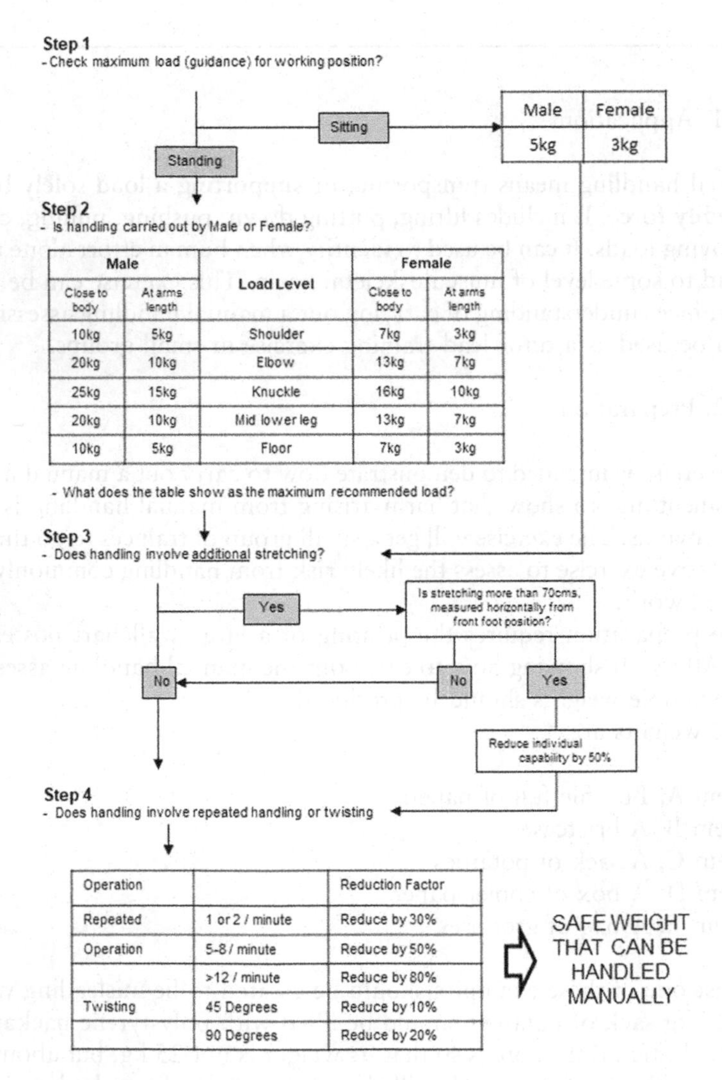
Manual Handling Exercise
Step 1
- Check maximum load (guidance) for working position?
Sitting
Standing
Step 2
- Is handling carried out by Male or Female?
- What does the table show as the maximum recommended load?
Step 3
- Does handling involve additional stretching?
Yes
No
Is stretching more than 70cms, measured horizontally from front foot position?
No
Yes
Reduce individual capability by 50%
Step 4
- Does handling involve repeated handling or twisting
SAFE WEIGHT THAT CAN BE HANDLED MANUALLY
Starting with the maximum recommended load for the person doing the handling calculate the recommended load allowing for reductions shown in steps 2,3 & 4

Male	Female
5kg	3kg

Male		Load Level	Female	
Close to body	At arms length		Close to body	At arms length
10kg	5kg	Shoulder	7kg	3kg
20kg	10kg	Elbow	13kg	7kg
25kg	15kg	Knuckle	16kg	10kg
20kg	10kg	Mid lower leg	13kg	7kg
10kg	5kg	Floor	7kg	3kg

Operation		Reduction Factor
Repeated	1 or 2 / minute	Reduce by 30%
Operation	5-8 / minute	Reduce by 50%
	>12 / minute	Reduce by 80%
Twisting	45 Degrees	Reduce by 10%
	90 Degrees	Reduce by 20%

Running the exercise:
The group are given a written brief as follows:

TASK 1 – Assessing weights
Estimate the weights (in kgs) of the items provided, without touching them.
Each member of the group should give their own estimate and then work out the average estimate (roughly). The final result should be entered on the Response Sheet under the "TASK 1" column.

Item A. Box file full of papers
Item B. A briefcase
Item C. A sack of potatoes
Item D. A box of copier paper
Item E. A bag of groceries

TASK 2 – How accurate were you?
Using the scales provided obtain and record the actual weight of the items A–E under the "TASK 2" column on the Response Sheet.

TASK 3 – Manual Handling Assessment
Using the wall flow-chart provided, assess which of the following tasks could lead to back injury: (i.e. Is the actual weight of the item being handled, heavier than the recommended maximum weight assessed from the chart?)

Operation 3a

A factory worker's task is to load sacks onto pallets (see photo 3a)

- *The operator is a reasonably fit male aged 38 years old*
- *Each pallet contains 24 bags*
- *It takes 25 minutes to fill the bags and stack one pallet*
- *The distance from the bag filler to the pallet is 3 metres*

Operation 3b

An office worker is lifting the box file from a shelf (see photo 3b)
The office worker is stretching from a sitting position

- *The action requires twisting through 60 degrees and the chair is not a swivel chair*
- *The task is done 20 times each day (10 times retrieving file and 10 times replacing file)*

Operation 3c

A sales manager keeps his briefcase in his car in the seatwell behind the drivers seat

- *The manager is a reasonably fit aged in his late 50s*
- *Access is restricted by adjacent vehicles*
- *The stretch distance inside the car is 50 cms*

TASK 4 – Controlling Risk
Identify which task is most likely to lead to an injury. Using the hierarchy of controls identify what actions could be taken to minimise the risk of injury occurring.

Produce your responses to each task on a flip chart and be prepared to report back to the other delegates.

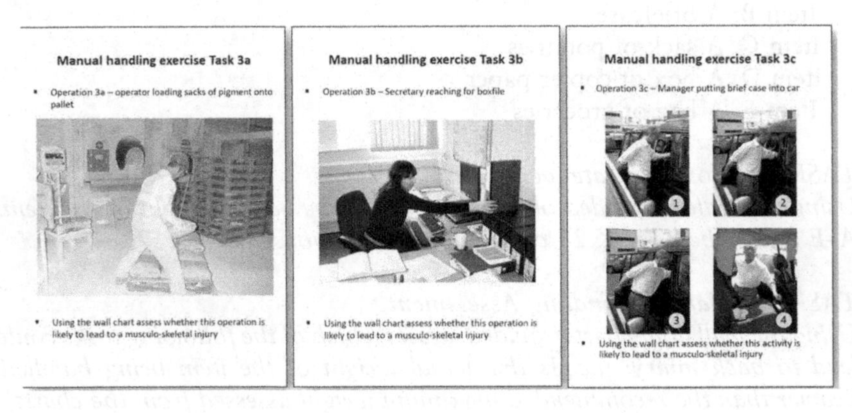

Running the Exercise:

This exercise takes 40 minutes to run, excluding any feedback time. It needs a full-time tutor to ensure that the exercise is explained and runs to time.

The tutor should explain the four parts of the exercise and then ask the trainees to make a guess at the weights of the five items. When guessing the weights, the trainees are not permitted to touch them. This should take about 5 minutes. When everyone has completed their estimates, these should be recorded as a single average result on the "Response Sheet".

The trainees should move onto task 2. This involves using luggage scales to weigh each item to get an accurate weight. In the interests of time, it is advisable for each trainee to be given a set of weigh scales. The average weight for each item is then recorded on the Response Sheet. The purpose of these two tasks is to demonstrate that a visual guess is not a reliable method of assessing the weights of items.

The tutor then explains Task 3. There are three scenarios shown on the exercise sheets 3a, 3b and 3c. Each of which shows a manual handling activity using one of the items which the teams have already weighed. The objective of this task is to assess which of the activities is most likely to lead to musculoskeletal harm. Before carrying out the assessment, the trainees are asked which they think is the highest risk activity. Typically, most people will guess that the loading of sacks onto a pallet is the highest risk. The carrying out of a structured manual handling risk assessment may change the trainees' views.

The tutor should demonstrate how to use the wall poster manual handling assessment flow chart by taking the trainees systematically through one of the scenarios. If the task 3b is taken as an example. Let us assume that the box file being handled by the office worker has been weighed at 3 kgs.

Using the flowchart then the maximum safe manual handling weight is found by the following steps.

Step 1 – The worker is in a sitting position and is female so immediately the maximum load is 3 kg.

Go straight to: Step 3 – The photograph shows that the task involves stretching so the allowable load must be reduced by 50% (i.e. from 3 kg to 1.5 kg).

Step 4 – There is no rapid repetition of the task, but the photograph shows twisting by about 45° so the allowable load must be reduced by a further 10% (i.e. from 1.5 kg to 1.35 kg).

Manual Handling Exercise

ITEM	LOAD		Guessed Weight (Team Average)	Actual Weight (Measured)	Assessed safe load	Order of Risk
A	Box File					
B	Briefcase					
C	Sack of Product					
D	Box of Copier Paper					
E	Bag of groceries					

Manual Handling Exercise Response Sheet

The assessment for exercise 3b shows that the safe load for the office worker to handle is 1.35 kg whereas the actual weight is 3 kg.

Having demonstrated how to carry out the manual handling assessment, the trainees are asked as a group to carry out the remaining two assessments themselves.

The final part of the exercise is task 4, where the trainees are asked to say what could be done to reduce the risk of injury in each of these three manual handling activities. A possible solution to this is to ensure that the worker stands up before lifting the file. Oddly the act of raising the height of the shelf might ensure that the worker always has to stand up to access this file without harm to her health.

Section S

Individual crosswords

Applications:

Crosswords can be used as a fun and non-threatening method of doing validations, gaining familiarity with new technical terms or introducing new safety slogans.

Of the two examples shown, the crossword style is best used for validation or new terminology and the pieceword can be used for introducing new safety campaign slogans. Both are useful in reinforcing safety messages in company newsletters or employee newspapers.

DOI: 10.1201/9781003342779-37

NOISE CROSSWORD

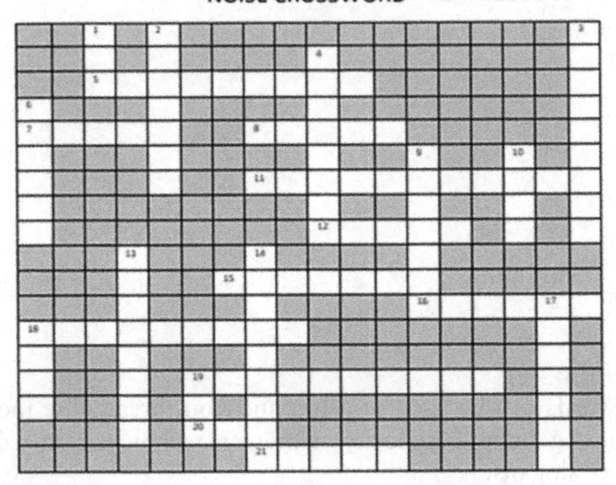

DOWN

1 Units of noise (absolute)
2 The spiral shaped cavity in the ear
3 The number of times that a sound occurs
4 Ringing in the ears
7 Bone in the middle ear (Linked to 6 across)
9 Able to be heard
11 Reduction in the amplitude of noise
12 Denoting or relating to sound
15 Unit of sound measurement
18 Unable to hear
19 A factor that causes deafness

ACROSS

5 Relating to sound
6 Bone in the middle ear (linked to 7 down)
8 External part of the human ear
10 Industrial deafness (initials)
13 Generic name for the bones of the middle ear
14 Hearing protection. Ear ___________(9)
16 First action level in decibels
17 Name of the ear drum
18 Wearable personal noise monitor
19 Medical test of hearing ability
20 Auditory organ
21 Vibrations that are capable of being heard

ANSWERS

		D		C													F
		b		O					T								R
		A	C	C	O	U	S	T	I	C							E
H				H					N								Q
A	N	V	I	L			P	I	N	N	A						U
M				E					I			A			N		E
M				A			A	T	T	E	N	U	A	T	I	O	N
E									U			D			H		C
R									S	O	N	I	C		L		Y
			O				D					B					
			S			D	E	C	I	B	E	L					
			S				F					E	I	G	H	T	Y
D	O	S	I	M	E	T	E	R								I	
E			C				N									M	
A			L		A	U	D	I	O	M	E	T	R	Y		P	
F			E		G		E									A	
			S		E	A	R									N	
							S	O	U	N	D					I	

Safety Pieceword

Rearrange the 3x3 pieces to form an important safety message

E	R	E
T	H	I
U	R	G

I	M	P
	T	H

	I	S
	G	E
H	U	R

O	R	T
A	T	
	W	O

A	N	T
I	T	

R	T	H
N	G	
F	O	R

	W	O
T	T	I
T		

	I	S
N	G	
E	N	T

	T	H
	N	O
S	O	

Solution

	T	H	E	R	E		I	S
	N	O	T	H	I	N	G	
S	O		U	R	G	E	N	T
			O	R				
I	M	P	O	R	T	A	N	T
	T	H	A	T		I	T	
	I	S		W	O	R	T	H
	G	E	T	T	I	N	G	
H	U	R	T			F	O	R

Section T

Personal commitment statement

T1. Application

Raise commitment to health and safety

This can be used at all levels in the organisation or just at the management level. Each individual is asked to make a personal written commitment to health and safety. The commitment text is usually produced through a consultation process, such as the organisation's Health and Safety Committee or representatives of the workforce. Each person is asked to sign the same agreed statement, which for managers might be something like:

My Personal Commitment to Health and Safety
I will... ....

- *Demonstrate my commitment to health and safety by my own personal example.*
- *Set clear health and safety expectations which are explained and verified for understanding and compliance on a regular basis.*
- *Focus on sustainable health and safety performance in all aspects of health and safety and will monitor progress regularly.*
- *Continually emphasise and demonstrate that production will not compromise health and safety.*
- *Consistently recognise good health and safety practice and performance, while confronting poor performance.*
- *Prioritise safety issues effectively and act quickly to resolve important issues.*

I will not ... ...

- *Fail to proactively manage safety (e.g. only acting when things go wrong).*
- *Delay following up on agreed safety actions.*

DOI: 10.1201/9781003342779-38

- *Tolerate variable and inconsistent health and safety standards.*
- *Allow short-term production pressures take priority over health and safety standards and workforce involvement.*

Signed ... . Date ... .

Section U

Hazard recognition

Application: Hazard recognition

Set up four different representations of hazards in each corner of the room. These might be:

i. A candle
ii. A bucket of water
iii. A chain saw
iv. A person (the tutor?)

(Note: Make sure that the items that you use do not create an actual hazard to the trainees!)

Ask the trainees to go and stand in the corner that represents the greatest hazard.

See how many people think that it is the "person" who could be the greatest hazard!

Discuss the reasons why they came to their conclusions.

DOI: 10.1201/9781003342779-39

Section V

Coincidence or not?

If:

If:

A	B	C	D	E	F	G	H	I	J	K	L	M	N	O	P	Q	R	S	T	U	V	W	X	Y	Z
1	2	3	4	5	6	7	8	9	10	11	12	13	14	15	16	17	18	19	20	21	22	23	24	25	26

Then **K - N - O - W - L - E - D - G - E**

11 + 14 + 15 + 23 + 12 + 5 + 4 + 7 + 5 = **96%**

and **H - A - R - D - W - O - R - K**

8 + 1 + 18 + 4 + 23 + 15 + 18 + 11 = **98%**

Both are important, but both fall short of 100%

BUT

A – T – T – I – T – U – D – E

1 + 20 + 20 + 9 + 20 + 21 + 4 + 5 = **100%**

Re-produced with thanks to Steve Fenton.
Safety is really about ATTITUDE. Make 100% safe behaviour your choice.

DOI: 10.1201/9781003342779-40

Section W

Unfamiliar task risk assessment slide rule

W1. Application

Training to assess and minimise risk when a task is only performed occasionally. This idea gets trainees involved in making their own personal risk assessment slide rule for use later "on the job".

W2. Preparation/Basis

The assessment is based on using a risk style matrix. It considers:

- How complex the job is
- How often this person has carried out the task
- The severity of the consequences if things go wrong

Basis for Unfamiliar Task Assessment system:

Task Complexity	Frequency performed	Minor Consequence (1)	Moderate Consequence (3)	Severe Consequence (6)
Straightforward (0)	Often	L (1)	L (3)	M (6)
	Seldom	L (2)	L (4)	M (7)
	Never	L (3)	M (5)	M (8)
Moderate (1)	Often	L (2)	M (4)	M (7)
	Seldom	L (3)	M (5)	M (8)
	Never	L (4)	M (6)	H (9)
Complex (2)	Often	L (3)	M (5)	M (8)
	Seldom	L (4)	M (6)	H (9)
	Never	M (5)	M (7)	H (10)

DOI: 10.1201/9781003342779-41

To use the assessment

- select the relevant complexity from the first column
- in that row select whether the task has been performed Often/Seldom/ Never
- consider the worst possible consequence and identify the column that corresponds to the appropriate consequence (Minor/Moderate/ Severe)
- The number in that cell shows the risk assessment rating

0–4 = Low (L-Green); 5–8 = Med (M-Amber); 9, 10 = High (H-Red)

W3. Converting the unfamiliar task assessment to a pocket slide rule.

A simple and practical way of carrying out the assessment at the point of work is to get the trainees to produce their own pocket slide rules.

The formats of the slide rule parts are shown in Fig W2 below. The parts page should be printed on card and laminated before being cut out and assembled. In this book the print is just black and white, but in making the calculator the following colours should be used:

i. Black shading = Red
ii. Medium grey = Yellow
iii. Light grey = Green

All five parts of the slide rule should be carefully cut out. The small window on the part bearing the name "STAMP" should be carefully cut out and then STAMP item should be folded and taped to form an open-topped long narrow pocket.

Strip "D" should be glued or taped to the back of strip "C", so that the coloured sides show on both the back and front of the glued strip.

The three strips can then be inserted in the pocket with the strip "A" first, then strip "B" and behind that strip "C", such that they all face towards the front (that features the word STAMP)

To use the slide rule:

1. Slide the green-topped strip (headed FREQUENCY) up to reveal the frequency that you have done this job before.
2. Holding the grey topped middle strip together with the green strip, slide both up to reveal the complexity of the task.
3. Finally holding all three strips together slide them up to reveal the severity of the consequences if things were to go wrong.
4. Check that each strip still shows the correct settings.

5. Turn the slide rule over – the colour in the small window shows the risk rating.
   - Green = proceed with care
   - Yellow = More precautions are necessary before proceeding
   - Red = STOP – do not go ahead. Consult with your boss.

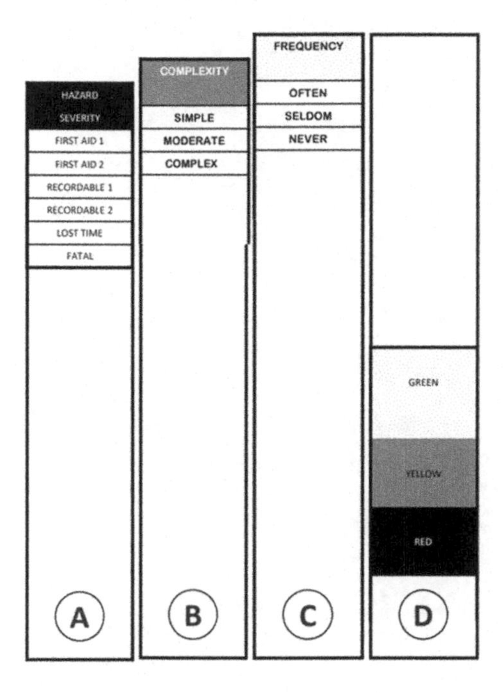

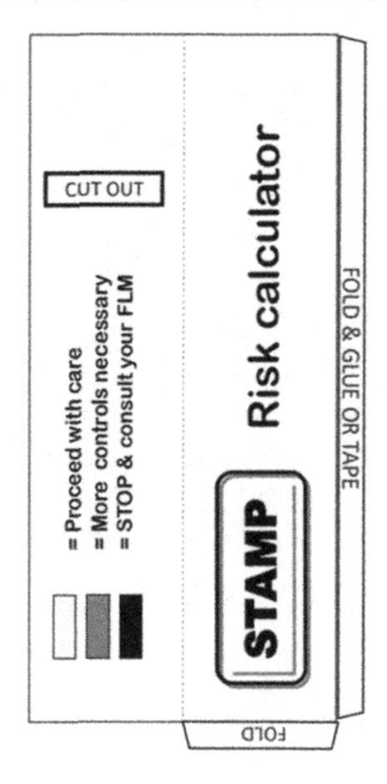

Section X

Microbooks

X1. Applications

Known under several tradenames, such as Microbooks, Z-fold Cards or Z-Folds, these relatively low-cost, credit card-sized documents are a great way of providing quite detailed training reminder information in a form that fits into workwear pockets, briefcases or handbags. They are ideal for providing health and safety information to visitors, truck drivers, business travellers and many other situations where handout materials are beneficial but space is short. They are particularly valuable where personal issues of emergency evacuation information may be needed. The microbook contains your bespoke printed image on a two-side printed thin paper which is folded in a concertina fashion such that it ends up credit card sized. This form of printing is available from many of the larger print rooms or from the internet.

The quality of these microbooks is such that they carry a much higher level of authority (they look "official") than normal photocopied or printed documents and because they are so easy to fit into a pocket, they can be easily carried around the workplace.

Getting representatives of the workforce involved in designing and preparing this type of handout material will encourage commitment and compliance.

DOI: 10.1201/9781003342779-42

Examples of microbooks:

Circle the hazard

Y1. Application: Homemade interactive exercises. Best used for individual trainees trying to spot the mistakes on a series of photographs, drawings or sketches.

For example: identifying hazards on a photograph of a workshop area, using a PowerPoint based display.

DOI: 10.1201/9781003342779-43

How to set up an interactive display in PowerPoint:

SLIDE 1

SLIDE 2

SLIDE 3

SLIDE 4

1. Decide how many hazards you will want to highlight during the interactive exercise.
2. Save the background slide as SLIDE 1 and duplicate two copies of the slide to produce the background for SLIDE 2 and SLIDE 3.
3. Locate the first hazard on SLIDE 2. Insert a shape to just cover the hazard (the hazard identifier shape). Set the hazard identifier shape to have no background, but temporarily give the shape a coloured border so that you can see it.
4. Insert a hyperlink on the hazard identifier shape to advance to SLIDE 3 when clicked. Hyperlinks are found on the Powerpoint toolbar by following the *Insert/Link/Place in this document.*
5. Insert a button (see bottom right corner in example) which says "CLICK HERE TO FIND ANOTHER HAZARD" and hyperlink that button to the first slide.
6. Duplicate SLIDE 3 enough times to reflect the number of hazards that the exercise will identify.
7. Return to SLIDE 3 and insert the symbol that you wish to identify a successful selection of a hazard. I use the explosion shape from the insert shapes tab using a bright colour and the text "HAZARD" (The shape should correspond to the position of the hazard identifier shape in SLIDE 2.)
8. You have now set up the first identifier. To add more hazards, go back to SLIDE 2 and position more transparent hazard identifier shapes by repeating steps 3–7. For each shape, locate a new explosion shape identifier on a new slide each time (see example for SLIDE 4

and onwards). There should never be more than one hazard identifier on the pages with blue buttons (SLIDES 3 onwards).

9. When all hazards have been included go back to SLIDE 2 and set all the hazard identifier shape border to "No Outline" so that they are invisible. Then delete SLIDE 1 and save the file. Once the interactive exercise works as you intended, then it is best saved as a Powerpoint show to stop trainees interfering with the structure!

Users find the hazards by clicking those areas on the photograph that they consider to be hazardous. It does not matter in which order they find the hazards.

A simple alternative to the electronic system is just to get trainees to circle the hazards on prints of the photograph or drawing. Humorous images can be obtained from several websites by searching "Spot the hazards cartoons".

Index

Note: *Italicized* and **bold** page numbers refer to figures and tables.

Printed in the United States
by Baker & Taylor Publisher Services